P9-CBK-908

AN INTRODUCTION TO VISCOUS FLOW

SERIES IN THERMAL AND FLUIDS ENGINEERING

JAMES P. HARTNETT and THOMAS F. IRVINE, JR., Editors
JACK P. HOLMAN, Senior Consulting Editor

Cebeci and Bradshaw • Momentum Transfer in Boundary Layers

Chang • Control of Flow Separation: Energy Conservation, Operational Efficiency, and Safety

Chi • Heat Pipe Theory and Practice: A Sourcebook

Eckert and Goldstein • Measurements in Heat Transfer, 2d edition

Edwards, Denny, and Mills • Transfer Processes: An Introduction to Diffusion, Convection, and Radiation

Fitch and Surjaatmadja • Introduction to Fluid Logic

Ginoux • Two-Phase Flows and Heat Transfer with Application to Nuclear Reactor Design Problems

Hsu and Graham • Transport Processes in Boiling and Two-Phase Systems, Including Near-Critical Fluids

Hughes • An Introduction to Viscous Flow

Kestin • A Course in Thermodynamics, revised printing

Kreith and Kreider • Principles of Solar Engineering

Lu • Introduction to the Mechanics of Viscous Fluids

Moore and Sieverding • Two-Phase Steam Flow in Turbines and Separators: Theory, Instrumentation, Engineering

Nogotov • Applications of Numerical Heat Transfer

Richards • Measurement of Unsteady Fluid Dynamic Phenomena

Sparrow and Cess • Radiation Heat Transfer, augmented edition

Tien and Lienhard • Statistical Thermodynamics, revised printing

Wirz and Smolderen • Numerical Methods in Fluid Dynamics

PROCEEDINGS

Hoogendoorn and Afgan • Energy Conservation in Heating, Cooling, and Ventilating Buildings: Heat and Mass Transfer Techniques and Alternatives

Keairns • Fluidization Technology

Spalding and Afgan • Heat Transfer and Turbulent Buoyant Convection: Studies and Applications for Natural Environment, Buildings, Engineering Systems

Zarić • Thermal Effluent Disposal from Power Generation

AN INTRODUCTION TO VISCOUS FLOW

WILLIAM F. HUGHES
Carnegie-Mellon University

● **HEMISPHERE PUBLISHING CORPORATION**

Washington　　　　London

McGRAW-HILL BOOK COMPANY

New York　　St. Louis　　San Francisco　　Auckland　　Bogotá
Düsseldorf　　Johannesburg　　London　　Madrid　　Mexico
Montreal　　New Delhi　　Panama　　Paris　　São Paulo
Singapore　　Sydney　　Tokyo　　Toronto

AN INTRODUCTION TO VISCOUS FLOW

Copyright © 1979 by Hemisphere Publishing Corporation. All rights reserved.
Printed in the United States of America. No part of this publication may be
reproduced, stored in a retrieval system, or transmitted, in any form or by
any means, electronic, mechanical, photocopying, recording, or otherwise,
without the prior written permission of the publisher.

1 2 3 4 5 6 7 8 9 0   K P K P   7 8 3 2 1 0 9 8

This book was set in Press Roman by Hemisphere Publishing Corporation.
The editors were Lynne Lackenbach and Christine Flint; the production
supervisor was Rebekah McKinney.
Kingsport Press, Inc., was printer and binder.

Library of Congress Cataloging in Publication Data

Hughes, William Frank, date
   An introduction to viscous flow.

   (Series in thermal and fluids engineering)
   Bibliography: p.
   Includes index.
   1. Viscous flow.   I. Title.
TA418.2.H83   620.1'06   78-14471
ISBN 0-07-031130-7

ENGR. LIBRARY

QA
929
.H76

CONTENTS

PREFACE

This book provides an introduction to viscous laminar flow, particularly boundary-layer theory, a subject of primary importance in fluid mechanics but seldom adequately treated at the elementary level.

As a textbook it should be suitable for a one-semester undergraduate course at the junior or senior level, or as the second half of a first course in fluid mechanics. It is assumed that the reader has had a brief introduction to the basic concepts of fluid mechanics—a course of the type usually presented at the sophomore level.

The approach is primarily physical, with emphasis on fluid behavior and engineering applications rather than on mathematical rigor. An elementary, physical approach to boundary layers is unique, the author believes, and readers at a more advanced level should also find it useful.

A few problems are included at the end of each chapter. They are not generally intended as exercises, but rather problems that illustrate applications and require some physical modeling. Often they are drawn from real engineering experience.

The author wishes to thank several persons for their help in the preparation of this book: Professor Chaim Gutfinger of the Technion, Haifa, Israel, for carefully reading the manuscript and making many useful suggestions; Professor Thomas F. Irvine, Jr., of the State University of New York at Stony Brook for the editorial tasks; and Miss E. Jean Stiles for patiently and skillfully typing the original manuscript.

William F. Hughes

AN INTRODUCTION
TO VISCOUS FLOW

ONE

INTRODUCTION

1-1 VISCOUS FLOW

Most of the matter in the universe is in a fluid state. Interstellar and intergalactic space is filled with gas, and the stars themselves consist of gases and liquid. Even the Earth, enveloped by a gaseous atmosphere, has a liquid core, and about two-thirds of its surface is covered by water. Indeed, the solid state is the unusual, not the common, state of matter in the universe.

It seems unnecessary, then, to justify the study of fluids. From an engineering point of view, which we shall try to maintain throughout this book, we can be almost as emphatic about the occurrence of fluids in problems of interest to the engineer. The fluids engineer needs to have a knowledge of the fields of aerodynamics, fluid machinery, fluid flow in pipes and conduits, and convective heat transfer, for example, and their application is necessary to the design of just about every device that moves or produces power.

Fluids, gases or liquids, may be classified according to their properties. In addition, a particular flow situation may be arbitrarily classified in many different ways depending on how the fluid behaves in that particular situation. For instance, we might say that the flow is steady or unsteady, compressible or incompressible, viscous or inviscid, laminar or turbulent, and so on.

In this book we are interested primarily in viscous flow as opposed to inviscid flow. As we shall show later (in detail in Chapter 5), a fluid flow field generally can be divided into two regions. In one region, away from a solid boundary, the fluid shear stress is unimportant and the flow is "inviscid." However, near a solid boundary where the fluid cannot slip along the surface, or in a pipe or conduit, shear stresses become important in establishing the flow dynamics.

These actual shear stresses in a fluid depend on the fluid "viscosity" if the flow is *laminar* or, if the fluid flow is turbulent, on the nature of the turbulence. In this book we shall be studying the flow in such regions where the shear stresses are important. We shall be interested mainly in *viscous* or *laminar* flow, not turbulent flow. The terms viscous flow and laminar flow are used interchangeably, and the term viscous laminar flow is redundant. On the other hand, under certain conditions (which we shall discuss later), viscous flow may become turbulent if it is speeded up. The study of turbulent flow is more complicated than the study of viscous flow. However, since turbulent flow is so important and occurs so often, we shall discuss some aspects of turbulent flow when necessary for completeness. In Chapters 5 and 6, when we study boundary-layer flow, we shall put turbulent flow on an equal footing with viscous flow.

Viscous flows are important in the design and analysis of many kinds of machinery and devices of interest to the engineer and in the study of natural phenomena of interest to the meteorologist, geophysicist, oceanographer, and astronomer.

In general, the determination as to whether a particular flow will be laminar or turbulent can be made by considering the numerical value of a dimensionless parameter of the flow, the Reynolds number, which we shall define and discuss in the next chapter. Usually, the flow changes from laminar to turbulent as the Reynolds number surpasses some critical value. For given fluid properties and flow configurations, the Reynolds number depends directly on the speed. Hence as the speed is increased the flow will change, sometimes rather suddenly, from laminar to turbulent.

Now, we shall review some of the fundamentals of fluid mechanics and discuss the physical basis for the concept of viscosity or resistance to shear in a moving fluid.

1-2 CONTROL VOLUMES AND SYSTEMS

There are two convenient viewpoints that one may take when formulating a quantitative description of the physical behavior of material media. The most convenient one to use in any particular situation depends on the problem. These two viewpoints are based on (1) a system approach, or (2) a control volume approach.

A system is an identifiable bit of matter, always consisting of the same elements or molecules. However, the system may move about, change shape, undergo thermodynamic property changes, or whatever. The essential thing is that the *material comprising a system is always the same* and the *mass of a system is constant.* In fact, a system may split up into several parts, the parts still forming the original system. A fixed group of particles (or molecules, say) may form a system regardless of how they move about. In particle dynamics, rigid-body dynamics, and thermodynamics, the concept of a system is useful.

On the other hand, a control volume is a much more useful concept for the description of fluid motion. A control volume is a *fixed, identifiable volume* in space. The volume is fixed in shape and size and usually is assumed fixed in an inertial coordinate system. This last assumption is not always necessary or convenient, but a control volume, by definition, is of a fixed shape (which of course is arbitrary and chosen to fit the problem). Hence material may flow into and out of a control volume. The mass of material in a control volume may change with time, and the properties of the mass in the control volume may change. When discussing what goes on in a control volume, we are actually discussing what goes on at a particular location in space and perhaps as a function of time—but the actual material may be continuously changing within the control volume.

By thinking about it a bit, one can understand why a control volume approach is better for the description of fluid flow. It would be virtually impossible to follow and keep track of every molecule of fluid, and even if we could the description would not be very useful. It makes much more sense to ask: What is going on at some particular position in space and time? That is, we would like to be able to "map" out the fluid motion with respect to some convenient coordinate system. For example, we would probably refer the velocity and flow rate in a river to some position on the bank.

These two viewpoints, system and control volume, lead to two different mathematical formulations of physical laws (see Fig. 1-1). The concept of a coordinate system has different meanings for a system approach or a control volume approach. If we were to keep track of a system (a particle, say), then a coordinate (x, say) could be used to denote the position of the particle with respect to some frame of reference. The motion would be given mathematically by an equation $x(t)$, where x is a dependent variable and time t is an independent variable.

In a control volume point of view, the spatial coordinates denote a position in space and we treat them (x, y, z, say) as independent variables along with t and describe the fluid motion and properties by equations such as $\mathbf{V} = \mathbf{V}(x, y, z, t)$, where $\mathbf{V}$ is the velocity vector. Similarly, we could have an equation for

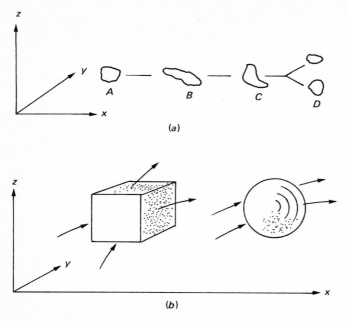

Figure 1-1 A system and a control volume. (*a*) A system consists of the same mass of material. It may move from point *A* to point *B* to point *C* and distort or actually break up into parts as shown at *D*. (*b*) Two control volumes, a cube and a sphere fixed in space relative to the co-ordinate system. Material (such as a fluid) may move through the control volume. A control volume can have any arbitrary shape we choose.

acceleration, thermodynamic properties, and so on, all functions of (x, y, z, t). This is the usual way one describes a fluid and the one we use throughout this book.

The manner of describing a fluid by a system or control volume in terms of the coordinates (x, y, z, t) with the meanings different for each is known as the Lagrangian or Eulerian description of motion, respectively. In the Eulerian description, which we use here, we always ask: What is the behavior or properties of a particular bit of fluid that happens to be at a certain prescribed place at a particular time; that is, we adopt a field theory point of view. As time goes on we are always looking at a different bit of fluid if we continue to focus our attention at the same position in space.

Of course, the actual numerical values for velocity, acceleration, thermo-dynamic properties, and so on, are identical in either coordinate system, but the mathematical formulation is different. An observer riding along with the fluid would measure the same numerical quantities as the observer fixed in space and using an Eulerian coordinate system if they measure them at the same time. As an example, consider a steady-state flow configuration as described by a Eulerian coordinate system. A motion picture of the flow would always look the same. At every point in space there is no change with time. As the fluid particles go by, each one looks and behaves just like its predecessor. However, an observer riding along with a fluid particle might see a change in time. Flow in a converging channel is a good example (Fig. 1-2). Consider a particle as it moves along the channel in steady flow. In steady flow the velocity at any position x does not change with time. Yet a particular particle changes velocity as it moves along. In Lagrangian coordinates the acceleration would be d^2x/dt^2, but in Eulerian coordinates we cannot differentiate x, and the velocity u (at a fixed point in space) is not a function of time. As we shall see later, the acceleration will be of the form $u(\partial u/\partial x)$ in Eulerian coordinates. In steady flow the numerical values of d^2x/dt^2 in Lagrangian coordinates and $u(\partial u/\partial x)$ in Eulerian coordinates are the same for this particular flow configuration.

1-3 THE BASIC LAWS OF FLUID MECHANICS

There are only a few very basic physical ideas from which the science of fluid mechanics is derived. Perhaps it is easiest to begin by expressing these fundamental concepts in words. The transition from words and ideas to a mathematical expression is not too difficult, but the mathematical form depends on whether we consider a system or control volume. In general, in fluid mechanics the control volume approach is simpler and more direct, and that is the approach we consider throughout this book.

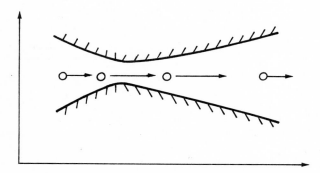

Figure 1-2 Flow in a channel and an area change, showing acceleration due to spatial effects. Even in steady flow there is acceleration. By continuity of flow, the velocity must be greater where the cross-sectional area is smaller.

There are three basic laws in fluid mechanics and indeed in mechanics in general. These laws take the form of statements about the conservation of mass, momentum, and energy. Although we shall not go into detail description here (it is assumed that the reader has some familiarity with these concepts), we shall briefly review the integral control volume forms of these three basic laws.

Conservation of Mass

The conservation of mass or continuity equation for a fluid states the following:

Rate of increase of material in the control volume = rate of net flux of material into the control volume

$$\frac{\partial}{\partial t} \int_{\text{vol}} \rho \, d(\text{vol}) = - \int_A \rho \mathbf{V} \cdot d\mathbf{A} \tag{1-1}$$

Here ρ is the mass density (slugs/ft^3 or kg/m^3), $\mathbf{V}$ is the velocity vector and $d\mathbf{A}$ is a vector of magnitude dA and direction normal (pointing outward) to the control volume. $\mathbf{V} \cdot d\mathbf{A}$ is the volumetric flow rate outward through the surface area dA. Physically, if at some location on the control volume surface the fluid is flowing outward, then the normal component of $\mathbf{V}$ is in the same direction as $d\mathbf{A}$ and $\mathbf{V} \cdot d\mathbf{A}$ is positive; if the flow is inward the value of $\mathbf{V} \cdot d\mathbf{A}$ is negative. The right-hand side of the above equation will be positive for net inward flow and will give rise to an increase in the mass within the control volume (as given by the left-hand side of the equation). The control volume is shown in Fig. 1-3.

The Momentum Balance or Equation of Motion

The momentum balance for a fluid is based on Newton's second law of motion. Newton's law in its basic form, force = (mass) × (acceleration), is valid for a constant-mass system. Hence the appropriate momentum balance for fluid in a fixed control volume must consider the momentum flux across the boundaries of the control volume. Again the reader is assumed to be familiar with the elementary derivation of the control volume form of this balance. The result is a vector balance among forces, momentum change inside the control volume, and momentum flux across the control surface:

Sum of the external forces acting on the fluid within the control volume = time rate of increase of the momentum within the control volume + net rate of momentum flux out of the control volume

$$\Sigma \mathbf{F} = \frac{\partial}{\partial t} \int_{\text{vol}} \rho \mathbf{V} \, d(\text{vol}) + \int_A \rho \mathbf{V}(\mathbf{V} \cdot d\mathbf{A}) \tag{1-2}$$

The forces **F** include all external forces, which in general consist of body forces such as gravity and surface forces such as those due to pressure and shear.

Conservation of Energy

We again consider an arbitrary control volume as shown in Fig. 1-4, but now considering energy flux crossing the boundary. We shall consider a "total" energy balance: Mechanical energy (i.e., kinetic and potential) and thermal energy (internal energy or enthalpy) will all be included, although as we shall see in Chapter 4, it is possible to separate the mechanical and thermal parts into separate balances. Again, it is assumed that the reader should recall from past introductory work in fluid mechanics that we may write the following balance for an arbitrary control volume. The time rate of increase of the total energy in the control volume equals the net rate of influx of total energy into the control volume plus the time rate of heat and work addition to the fluid in the control

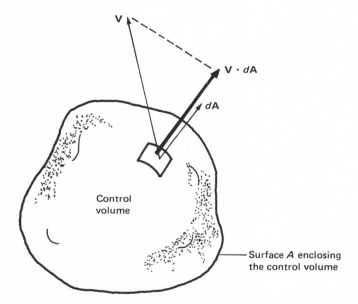

Figure 1-3 A general control volume showing an element $d\mathbf{A}$ and velocity vector **V**. The term **V** $\cdot$ $d\mathbf{A}$ is the component of **V** normal to the surface multiplied by the area dA and physically represents the volumetric flow rate outward through the area dA. The term $\rho\mathbf{V}(\mathbf{V} \cdot d\mathbf{A})$ then is a vector parallel to **V** and represents the (scalar) mass flow rate through dA multiplied by the velocity vector **V** and physically is simply the rate of momentum flux through the elemental area dA.

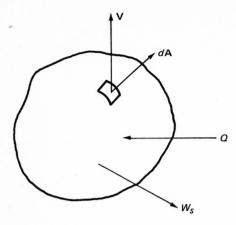

Figure 1-4 A control volume energy balance.

volume. The specific energy density (density per unit mass) is $e + (V^2/2) + \psi$, where e is the specific internal energy, $V^2/2$ the specific kinetic energy, and ψ the specific potential energy.*

Time rate of increase of energy in the control volume $=$ time rate of influx of energy into the control volume + rate of heat addition to the fluid in the control volume + rate of work done by the pressure forces at the boundary on the fluid in the control volume − rate of shaft work or irreversible work done by the fluid in the control volume

$$\frac{\partial}{\partial t} \int_{\text{vol}} \rho \left(e + \frac{V^2}{2} + \psi \right) d(\text{vol}) = - \int_A \rho \left(e + \frac{V^2}{2} + \psi \right) \mathbf{V} \cdot d\mathbf{A} + Q$$

$$+ \left(- \int_A P \mathbf{V} \cdot d\mathbf{A} \right) - W_s$$

Condensing together, we get the familiar form of the control volume energy equation:

$$\frac{\partial}{\partial t} \int_{\text{vol}} \rho \left(e + \frac{V^2}{2} + \psi \right) d(\text{vol}) + \int_A \rho \left(e + \frac{P}{\rho} + \frac{V^2}{2} + \psi \right) V \cdot dA = Q - W_s \qquad (1\text{-}3)$$

---

*For example, if z is the elevation above some arbitrary datum level, $\psi = gz$, the gravitational potential, and the force due to gravity is $-\partial \psi / \partial z = -g$ per unit mass or $-\rho g$ per unit volume.

We shall have more to say about this equation later on when we use it to begin our development of the general energy equation in differential form.

1-4 THE CONCEPT OF VISCOSITY

Viscosity is a measure of the magnitude of the shear stress generated in a fluid when it is in motion. Put another way, viscosity is a measure of a fluid's ability to resist motion when a shearing stress is applied. One might visualize the concept of viscosity as follows. Whenever a fluid is in motion and is being sheared, we can imagine the fluid to be made up of a very large number of layers, each with a different velocity as shown in Fig. 1-5. Consider two adjacent layers. The faster layer tends to pull the slower layer exerting a stress τ on it and urging it on and the slower layer tends to retard the faster layer.

In general the shear stress τ in a fluid depends on the state of motion of the fluid and on certain physical properties of the fluid. The simplest type of relationship and the one that proves to be the most useful—good for gases, water, oil, and indeed most but not all liquids—is the Newtonian relationship. Fluids that are described by this relationship are known as Newtonian fluids. Newtonian fluids are ones in which the shear stresses are *linearly* related to the strain rates (which can be expressed in terms of velocity gradients). As an example, let us look more closely at a particularly simple shear flow.

Consider the laminar viscous flow of fluid between two parallel infinite plates. The top plate moves with velocity V relative to the bottom plate as shown in Fig. 1-6. At the surface of each plate the fluid moves with the plate and does not slip. This "no-slip" condition is a fundamental concept and is universally true in all fluid flow except for very rarefied gas flow, which we shall not consider in this book. The velocity profile will be a straight line between the two plates (as we shall presently prove). We show free bodies of the plates and

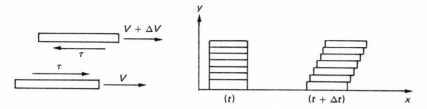

Figure 1-5 The top layer moving at velocity $V + \Delta V$ tends to pull along the bottom layer moving slower at velocity V. The shear stress τ is exerted one on the other. A stack of layers will move as shown in a time Δt.

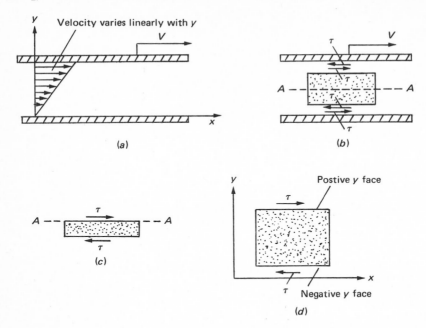

Figure 1-6 The shear stress in a simple shear flow configuration. (*a*) shows the velocity profile. (*b*) is a free body of the fluid contained in a fixed control volume between the plate. (*c*) shows a free body of an arbitrary portion of the control volume to illustrate that τ is uniform throughout such a flow. (*d*) illustrates the definition of a positive and negative face.

the fluid between the plates at some instant of time. In steady state the magnitude of the shear stress τ is the same on the top and bottom plate.

The directions of τ shown are according to the convention that on a positive y face τ is measured positive in the positive x direction and on the negative y face τ is positive in the negative x direction. Now, if we imagine a cut A-A through the fluid, a free body of the bottom portion is shown in Fig. 1-6(*c*). Since this portion is in equilibrium, the magnitude of shear stress on the surface A-A must be equal to that on the bottom. Hence τ is uniform throughout the fluid.

Now, the Newtonian hypothesis states that this shear stress τ may be expressed as

$$\tau = \mu\left(\frac{V}{h}\right)$$

where μ is a property of the fluid known as viscosity. (For a non-Newtonian fluid such as grease or a polymer, say, which we shall discuss later, the relationship between τ and the kinematics of flow may be much more complex.) The above Newtonian expression can be easily generalized to a one-dimensional flow where $u = u(y)$ as shown in Fig. 1-7. Locally τ is proportional to $\partial u/\partial y$ and

$$\tau = \mu \frac{\partial u}{\partial y} \tag{1-4}$$

If we remove a layer of fluid in this flow configuration, obviously the shear stress is not the same on the top as the bottom since $\partial u/\partial y$ is not uniform with y. But since the element is in equilibrium, the difference in τ is balanced by the pressure differential on the x faces. We shall examine this problem in detail in the next chapter, where we shall discuss a more general Newtonian relationship for arbitrary three-dimensional flow.

The units of viscosity are indicated by the relationship

$$\mu = \frac{\tau}{\partial u/\partial y}$$

In English engineering units the units of μ are $lb_f \cdot s/ft^2$, and in SI units $N \cdot s/m^2$ or $Pa \cdot s$. Several other units are in use. Of particular interest are the cgs units, in which τ is measured in dyn/cm^2 and u in cm/s. Then μ is given in $dyn \cdot s/cm^2$. A $dyn \cdot s/cm^2$ is known as a *poise*, and is used in some scientific work but rarely in engineering.

The ratio μ/ρ occurs quite frequently in fluid mechanics; it is known as kinematic viscosity and is generally denoted as ν. In the English engineering system ρ has units of $slugs/ft^3$ and ν has units of ft^2/s. In the SI system ρ has units of kg/m^3 and ν has units m^2/s. Again in the cgs system, if ρ is given in g/cm^3, then ν is in units of cm^2/s. A cm^2/s is known as a *stoke*; again, it is seldom used any more in engineering calculations but is used extensively in scientific work.

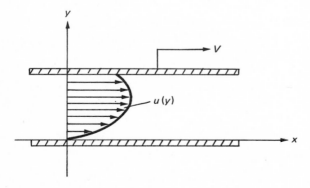

Figure 1-7 A generalized one-dimensional laminar flow.

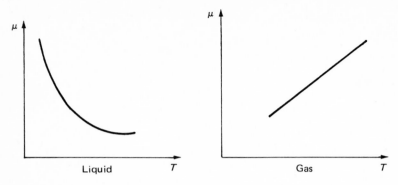

Figure 1-8 Viscosity as a function of temperature, T.

In summary, viscosity is a fluid property that indicates the fluid's resistance to shearing motion. The viscosity does not depend on density and, indeed, fluids less dense than water, say, may have a higher viscosity—for example, oil. The common notion of a "heavy" oil refers to the large viscosity of the oil, not its density. Viscosity depends strongly on temperature and to some extent on pressure. The viscosity of a liquid decreases with increasing temperature and the viscosity of a gas does just the opposite, increasing with increasing temperature (Fig. 1-8).

1-5 UNITS IN FLUID MECHANICS

The mathematical form of all the equations and laws that will be developed in this book do not depend on the units to be used. We can and shall use two different systems of units—the English engineering system and the metric SI system. Eventually the SI system (International Scientific or *Le Systeme International d'Unites*) will be used universally throughout the world, but at present the English engineering system is still in widespread use in the United States.

In the Appendix a more complete discussion of units is presented along with tables of units for different quantities and conversion factors.

There are two ways to obtain a consistent set of units in any given system. Either M, L, T, θ (mass, length, time, angle) or F, L, T, θ (force, length, time, angle) may be used and all other dimensions expressed in terms of these chosen independent dimensions by means of laws and definitions.

The choice of beginning with F, L, T, θ or M, L, T, θ is arbitrary, but the units of force and mass are related in a definite way. We can begin with Newton's second law of motion,

Table 1-1 Dimensions and Units

	English Engineering	SI
Force	pound (lb)	newton (N)
Mass	slug	kilogram (kg)
Time	second (s)	second (s)
Length	foot (ft)	meter (m)
Pressure	lb/ft²	pascal = N/m² = (Pa)[a]
Velocity	ft/s	m/s
Viscosity	lb · s/ft²	N · s/m² = Pa · s
Temperature	degree Rankin (°R)	Kelvin (K)

[a]The kilopascal (kPa) is frequently used because the pascal is such a small unit.

$$\mathbf{F} = m\mathbf{a}$$

and if we choose the units for force, the units of mass follow directly. The unit of mass is the amount of mass that would be accelerated at the rate of a unit distance per square second when acted upon by a unit force. In English units the unit of force is the pound force (lb_f) and the unit of mass is the amount that is accelerated 1 ft/s² when acted upon by 1 lb_f. This unit of mass is called a slug. In the SI system the unit of force is the newton (N) and the unit of mass is the kilogram (kg). A 1-kg mass will be accelerated 1 m/s² when acted upon by 1 N.

There are other ways to relate mass and force independently of Newton's law. The pound mass (lb_m) may be defined as the amount of mass that would be attracted toward the Earth's surface by a force of 1 lb. When one defines units of force and mass independent of Newton's law, a conversion factor g_c must be introduced to ensure dimensional homogeneity when Newton's law is written (after the fact, so to speak). A more detailed discussion of g_c is given in the Appendix.

However, we shall be using Newton's law of motion to relate mass and force throughout this book, hence eliminating the need for the so-called g_c conversion factor, which seems to add more confusion than not. Now, in our system, the magnitude of the force due to gravity acting on a mass is always simply Mg, where g is the local acceleration of gravity. Mg is then the "weight" of a mass M.

Table 1-1 is a brief table of units of some quantities that appear frequently in fluid mechanics. A more complete listing and discussion of units are presented in Appendix C.

TWO

VISCOUS INCOMPRESSIBLE FLOW

2-1 INTRODUCTION

Fluids often flow in a laminar manner, although an observer making a casual observation of his surroundings might think that fluids nearly always flow turbulently. When fluids move fast or on a large scale, they do indeed tend to flow in a turbulent manner, but nevertheless the study of laminar viscous flow is of vital importance in engineering and the study of viscous flow is not only useful in itself but forms the basis for the study of turbulent flow.

Some of the applications of viscous flow theory that we shall discuss are to hydrodynamic lubrication, low-speed flow in pipes and channels, and boundary-layer theory. Bearings are designed so that the lubricant flow is laminar, and in all boundary layers the flow is laminar over at least a portion of the extent of the boundary layer. We shall discuss these applications in detail later, but first let us focus our attention on the simplest type of viscous flow, the flow between parallel plates and in pipes, examples of simple shear flows. After that we shall derive the general equations of continuity and motion that apply to all fluids in viscous flow.

When studying a fluid that is assumed to be incompressible, we need not use the energy equation to describe the flow. The motion or momentum equation (which is a vector equation with three components) and the continuity equation (which is a scalar equation) suffice to determine, at least in theory, the velocity components and pressure throughout the fluid. The flow of heat in an incompressible fluid is usually uncoupled from the dynamics of the flow so that the heat flow, which must be found from the energy equation, can be formulated as a separate problem after the dynamics are studied. In an incompressible fluid the internal friction that comes about because of viscosity is a mechanical dissipation, which may remain in the fluid, increasing its internal energy and temperature, or may be transferred out through heat conduction. This entire balance is described by the fluid energy equation and is a separate problem from the dynamics, which determine the velocity and pressure in the fluid. However, things are a bit more complicated if, for example, the viscosity variation with temperature is considered. Then the equation of motion and the energy equation are coupled together and must be solved simultaneously (along with the continuity equation), since the temperature (an unknown quantity, which might vary throughout the fluid) appears in both equations. In this book, however, we shall not consider viscosity variations and can ignore the energy equation for the moment while we examine the equation of motion. Later, we shall consider the energy equation but shall be able to treat its solution as a separate problem.

Of course, if we were studying compressible fluids, we would need the energy equation and perhaps an equation of state and would need to have all four equations, motion, continuity, energy, and state, in the mathematical formulation of general compressible fluid flow. The reason why we always need an energy equation (or an equivalent relationship) is that the density appears in the equation of motion, and this is a variable that depends not only on pressure, but on temperature as well. However, viscous compressible flow is a rather complicated subject, and we shall confine ourselves mainly to incompressible flow in this book. Remember, however, that gases may be treated as incompressible fluids under certain circumstances (roughly when the speed is slow, Mach number less than about 0.3, and the external heating or cooling is small) and the results of incompressible flow are applied with good accuracy even to subsonic aerodynamics.

Hence, the study of incompressible viscous flow is not really so restrictive as it might seem, and indeed forms an essential part of modern fluid dynamics with applications ranging from oceanography to lubrication to aerodynamics.

2-2 SIMPLE VISCOUS SHEAR FLOW

We begin our study of viscous flow by considering steady flow in a simple geometry in which the velocity varies in only one dimension, the flow between two very large (essentially infinite) parallel plates. If both plates are stationary, the flow is known as *Poiseuille flow*, and there must be a pressure variation (or gradient) in the direction of flow in order to force the fluid to move. If one plate slides with respect to the other fixed plate in the direction of flow, the flow is known as *Couette flow*, and a pressure variation may or may not be present. In Fig. 2-1 these two configurations are shown together with the velocity profile (the velocity variation across the flow channel). As we shall see, the profile for Poiseuille flow is parabolic, and for Couette flow the profile is a straight line if there is no pressure gradient in the direction of flow. If there is a pressure gradient, the profile is the superposition of a straight line and a parabola, the parabola adding or subtracting to the straight line depending on the sign of the pressure gradient, that is, the direction in which the pressure increases. In other words, the pressure forces the fluid in one direction and the motion of the top plate can aid, retard, or override in the opposite direction. These possibilities are shown in Fig. 2-1*a* through 2-1*e*.

In reality, of course, the plates must have a beginning and end in the direction of flow. At the entrance of the channel the fluid must enter the space between the plates with some initial velocity profile. The simplest profile is a uniform velocity distribution and is the one that occurs frequently in practice.

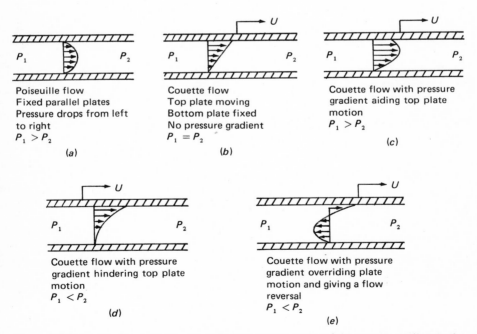

Poiseuille flow
Fixed parallel plates
Pressure drops from left
to right
$P_1 > P_2$

(a)

Couette flow
Top plate moving
Bottom plate fixed
No pressure gradient
$P_1 = P_2$

(b)

Couette flow with pressure
gradient aiding top plate
motion
$P_1 > P_2$

(c)

Couette flow with pressure
gradient hindering top plate
motion
$P_1 < P_2$

(d)

Couette flow with pressure
gradient overriding plate
motion and giving a flow
reversal
$P_1 < P_2$

(e)

Figure 2-1 Poiseuille flow and Couette flow with various values of the pressure gradient in the direction of flow.

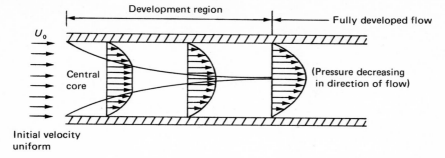

Figure 2-2 The development region in Poiseuille flow. The effects of viscosity move toward the center of the channel and eventually form the fully developed profile.

When the fluid enters the channel, the velocity of the fluid at the walls must be zero relative to the wall because of the no-slip condition, and the profile begins to distort and develop into the appropriate profile shown in Fig. 2-1, which then does not vary on down the channel. This initial region is known as the *development region* and may extend a distance equal to several times the spacing between the plates, as shown in Fig. 2-2. However, the behavior in this region is rather complicated, and we shall reserve discussion of it until we study boundary layers in a later chapter. If the plates are long compared to the spacing, we can ignore the development region. This is a valid procedure in many engineering problems, such as in lubrication theory, where the spacing between plates is only a few thousandths of an inch and the length several inches.

We shall assume that the flow is laminar, steady, and incompressible and find the velocity profile and pressure variation in these basic configurations. Remember that if the Reynolds number, $\rho h U / \mu$ is too large, then the flow will become turbulent. Here h is the spacing between plates and U is the velocity of the top plate or mean velocity of the flow.

Let us begin by extracting a small elemental control volume and writing a momentum-force balance on the fluid contained in it to obtain an equation of motion in differential form. Consider an element Δx by Δy by a unit dimension out of the paper as the control volume. Figure 2-3 shows the element and coordinate system and forces acting. The origin is arbitrary, but we can take it at the entrance to the channel if we neglect entrance or development effects. Since the flow is fully developed, the velocity in the x direction, which we denote as u, will be a function of y only, and hence the momentum flux across the vertical surface at x will be the same as at the vertical surface at $x + \Delta x$ and will not enter into the equation of motion. There will be pressure forces on the two vertical faces as shown and shear forces on the top and bottom faces. For a

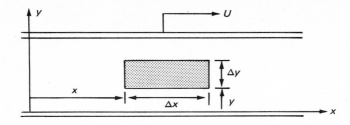

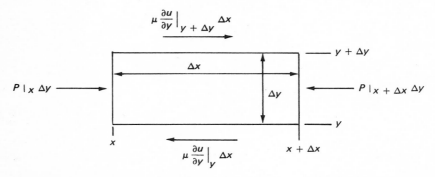

Figure 2-3 An elemental control volume for a momentum-force balance showing the forces acting. The forces in the x direction are shown.

Newtonian fluid, remember that the shear stress is proportional to the velocity gradient $\partial u/\partial y$ and hence on the top face the shear stress in the x direction is $\mu(\partial u/\partial y)$ evaluated at $y + \Delta y$ and on the bottom face the shear in the negative x direction is $\mu(\partial u/\partial y)$ evaluated at y. These expressions for the shear are correct regardless of the sign of $\partial u/\partial y$ or the sign of u. A negative value of $\partial u/\partial y$ would give a negative shear, and so on. The significance of the expressions may be realized intuitively by imagining a flow in which u increases in the positive y direction (positive $\partial u/\partial y$) and imagining the direction of the shear being exerted on the element by the fluid on top and bottom. The fluid on top tends to move faster and drag the fluid with it, and the fluid on bottom tends to retard the fluid in the element. Only the forces in the x direction are shown. Actually there are shears in the y direction on the vertical faces and pressure forces on the horizontal faces, but as we shall see, there is no motion in the y direction (in fully developed flow) and the y component of the equation of motion is unnecessary. There may be a hydrostatic pressure variation in the y direction, but this is irrelevant now and generates no motion and can always be added to the pressure later.

Now we write the steady-state force balance as

$$F_x = \int_A \rho u \mathbf{V} \cdot d\mathbf{A} = 0$$

$$P|_x \, \Delta y - P|_{x + \Delta x} \, \Delta y + \mu \left.\frac{\partial u}{\partial y}\right|_{y + \Delta y} \Delta x - \mu \left.\frac{\partial u}{\partial y}\right|_y \Delta x = 0 \qquad (2\text{-}1)$$

where F_x is the sum of the external forces including stresses acting in the x direction and $\mathbf{V}$ is the velocity vector.

Now we divide through by $\Delta x \, \Delta y$ and take the limit as $\Delta x \to 0$ and $\Delta y \to 0$.

$$- \lim_{\Delta x \to 0} \frac{P|_{x + \Delta x} - P|_x}{\Delta x} + \lim_{\Delta y \to 0} \frac{\mu \left.\frac{\partial u}{\partial y}\right|_{y + \Delta y} - \mu \left.\frac{\partial u}{\partial y}\right|_y}{\Delta y} = 0$$

These terms are precisely the definitions of partial derivatives and hence obtain

$$0 = -\frac{\partial P}{\partial x} + \mu \frac{\partial^2 u}{\partial y^2} \qquad (2\text{-}2)$$

which is a very basic equation of one-dimensional viscous flow, which we will be using extensively. The reason for writing the zero on the left is convention, and later on we shall place inertia terms arising from the momentum flux on the left-hand side. We have assumed the viscosity μ to be constant, as we shall throughout this book. However, even for variable viscosity (with temperature and pressure), μ may be brought outside the derivative to a high degree of accuracy and Eq. (2-2) used.

Before solving Eq. (2-2), we can make some observations about the nature of the solution. In fully developed flow for which Eq. (2-2) holds, u is a function of y only and hence $\partial^2 u/\partial y^2$ is a function of y only. Hence $\partial P/\partial x$ must be either a constant or a function of y and not a function of x. We can easily show that $\partial P/\partial x$ must in fact be a constant. If we neglect any hydrostatic effects and assume that P is not a function of y, then $\partial P/\partial x$ must be a constant, obviously. Furthermore, even if gravity effects generate a hydrostatic pressure, the value of $\partial P/\partial y$ is a constant and

$$\frac{\partial}{\partial x}\left(\frac{\partial P}{\partial y}\right) = 0$$

and interchanging the order of differentiation, we have

$$\frac{\partial}{\partial y}\left(\frac{\partial P}{\partial x}\right) = 0$$

and hence $\partial P/\partial x$ is a function of neither x nor y and must be a constant. The value of $\partial P/\partial y$ may be found by writing the y component of the equation of motion. We can redraw Fig. 2-3 showing only the y force components, assuming gravity to act in the negative y direction (Fig. 2-4). Following the previous procedure, we find that

$$0 = -\frac{\partial P}{\partial y} - \rho g \qquad (2\text{-}3)$$

The two shears cancel out as the limiting process is taken. The presence of these shears on vertical faces may seem strange, but they are necessary to ensure rotational equilibrium of the element of fluid. We shall examine these stresses in more detail when we discuss kinematics of fluids in a later section.

We shall now solve Eq. (2-2) for various boundary conditions.

2-3 POISEUILLE AND COUETTE FLOW

We now apply Eq. (2-2) to the viscous steady flow of an incompressible fluid between parallel plates. We begin by assuming both plates stationary. Referring to Fig. 2-5, the plates are spaced a distance h apart, are of length L, and of very wide width out of the paper such that this width is very large compared to h and

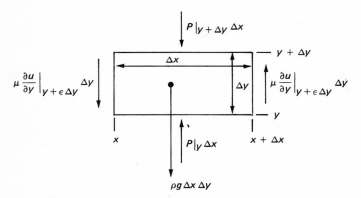

Figure 2-4 The y components of force in fully developed viscous flow. Gravity is assumed to be in the negative y direction. The shears generate torques that just balance those shown in Fig. 2-3. The vertical shears are evaluated at some mean position to ensure rotational equilibrium of the element.

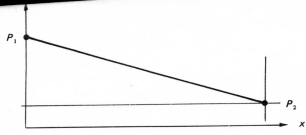

Figure 2-5 Poiseuille flow between parallel plates. The pressure distribution is a straight line between P_1 and P_2.

L. We neglect the entrance region and look only at the fully developed flow, which is equivalent to assuming that $h \ll L$. The pressure at the inlet, $x = 0$, is P_1, and at the outlet, $x = L$, the pressure is P_2 (see Fig. 2-5). Pressure variations in the y direction are at most hydrostatic and not considered here.

The boundary conditions on the velocity are that there is no slip at the walls. Hence we must solve

$$0 = -\frac{\partial P}{\partial x} + \mu \frac{\partial^2 u}{\partial y^2}$$

with the conditions

$$u = 0 \quad y = 0, h \tag{2-4}$$

Now, since u is a function of y only and $\partial P/\partial x$ is a constant, we can write the velocity derivative as an ordinary derivative, which can be integrated directly to find $u(y)$.

$$\frac{d^2 u}{dy^2} = \frac{1}{\mu} \frac{\partial P}{\partial x} \tag{2-5}$$

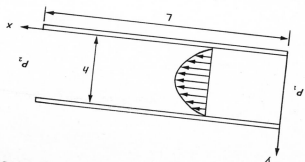

equation. Den~~...~~

channel as Q, we have

$$Q = \int_0^h u \, dy = -\frac{h^3}{12\mu} \frac{\partial P}{\partial x} \tag{2-9}$$

The negative sign means that if the flow Q is positive (in the positive x direction), then, $\partial P/\partial x$ must be negative; that is, the pressure must decrease in the positive x direction. Intuitively this is reasonable, since the pressure must be larger at the pipe inlet to push the fluid along. If there is no pressure gradient, there is no flow. Since $\partial P/\partial x$ is constant, the pressure variation must be a straight line between P_1 and P_2 and we can replace $\partial P/\partial x$ by $(P_2 - P_1)/L$. The flow rate then is simply

$$Q = \frac{h^3(P_1 - P_2)}{12\mu L} \tag{2-10}$$

which gives the flow in terms of the difference in pressure across the channel.

Here Q is the flow rate per unit width of the channel and has dimensions of ft^2/s or m^2/s, say. The average velocity $\bar{U}$ over the channel is defined by

$$\bar{U} = \frac{\displaystyle\int_0^h u\,dy}{\displaystyle\int_0^h dy} = \frac{1}{h}\int_0^h u\,dy = \frac{Q}{h}$$

We now extend the analysis to the case where the plates are in relative motion. This type of flow is known as Couette flow. In Fig. 2-6 the top plate moves with velocity U relative to the fixed plate of length L. The pressures are, as before, P_1 and P_2 at the inlet and outlet of the flow channel, respectively. As before, we solve Eq. (2-2), but now the boundary conditions are

$$\begin{aligned} u &= 0 & y &= 0 \\ u &= U & y &= h \end{aligned} \qquad (2\text{-}11)$$

Applying these to Eq. (2-6), we have

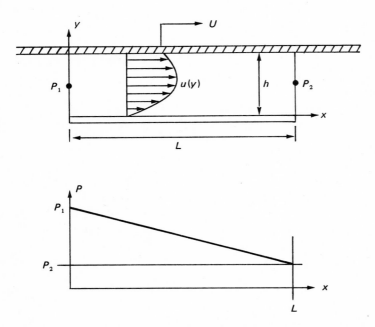

Figure 2-6 Couette flow between parallel plates. The coordinate system is attached to the bottom fixed plate.

$$U = \frac{1}{\mu} \frac{\partial P}{\partial x} \frac{h^2}{2} + C_1 h$$

$$0 = C_2$$

so that

$$C_1 = \frac{U}{h} - \frac{1}{\mu} \frac{\partial P}{\partial x} \frac{h}{2} \tag{2-12}$$

$$C_2 = 0$$

and the velocity profile is

$$u = \frac{1}{2\mu} \frac{\partial P}{\partial x}(y^2 - hy) + \frac{Uy}{h} \tag{2-13}$$

This is just like Eq. (2-8) with an additional term Uy/h, which comes about because of the motion of the top plate. The effects of pressure and plate motion are simply linearly superimposed. If there were no pressure gradient, then the velocity would be simply $u = Uy/h$, a straight line between the plates. This would then be a simple shear flow with uniform shear throughout the fluid.

Depending on the sign of $\partial P/\partial x$, the parabolic term may add or subtract to the linear term and generate profiles not only like that of Fig. 2-6 but also like those shown in Fig. 2-1.

The flow rate Q is (with respect to the fixed plate)

$$Q = \int_0^h u \, dy = -\frac{h^3}{12\mu} \frac{\partial P}{\partial x} + \frac{Uh}{2} \tag{2-14}$$

again merely the superposition of Poiseuille flow and simple shear flow. The pressure gradient is again a constant, and we can express the flow rate as

$$Q = \frac{h^3(P_1 - P_2)}{12\mu L} + \frac{Uh}{2} \tag{2-15}$$

The shear stress τ anywhere in the fluid is simply $\mu(\partial u/\partial y)$. At the walls the shear stress on the wall, τ_w, is $\mu(\partial u/\partial y)|_{y=0}$ on the bottom wall and $-\mu(\partial u/\partial y)|_{y=h}$ on the top wall, the shear stress on the wall being defined positive in the positive x direction. On an element of fluid the shears are as shown in Fig. 2-3.

2-4 FLOW IN RECTANGULAR CHANNELS

We have assumed so far that the plates were very wide. If this is not the case and we have instead a rectangular channel, the problem becomes more complex and

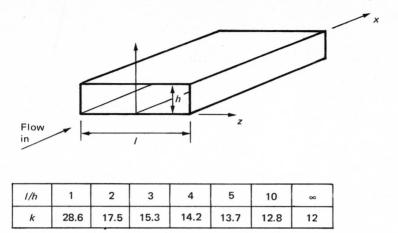

l/h	1	2	3	4	5	10	∞
k	28.6	17.5	15.3	14.2	13.7	12.8	12

Figure 2-7 The rectangular-cross-section channel and values of k, the proportionality constant between pressure gradient and flow rate.

the velocity u is a function not only of y but also of another coordinate z. Then the equation of motion becomes a partial differential equation with a Fourier series solution. We shall not discuss this problem here in detail, but some of the results will be stated for reference and comparison. Figure 2-7 shows the rectangular channel.

It is convenient to introduce the average velocity $\bar{U}$ in the rectangular channel as

$$\bar{U} = \frac{\displaystyle\int u \, dA}{\displaystyle\int dA} = \frac{1}{A} \int u \, dA = \frac{Q}{A}$$

Here Q is the total volumetric flow rate in the channel with units of ft^3/s or m^3/s. We relate the pressure gradient to $\bar{U}$ as

$$\frac{\partial P}{\partial x} = -\frac{k\mu\bar{U}}{h^2} \tag{2-16}$$

For very wide plates ($l \gg h$), we see from Equation (2-9) that k is 12. Figure 2-7 includes a table of k versus the aspect ratio (l/h), which allows the flow rate to be found for rectangular channels although we shall not find the velocity profile here. As before, the pressure variation is just a straight line.

2-5 FLOW IN CIRCULAR PIPES

In practice, most pipes and channels for liquid flows are circular, and Poiseuille flows in pipes form the starting point of the plumbing industry. We begin by deriving a differential equation for steady laminar flow. As before, we take out a small elemental control volume, but now we work in cylindrical coordinates. It is convenient to consider an annular cylinder of thickness Δr and length Δx (Fig. 2-8a) in a pipe of radius R.

Since the flow is assumed to be fully developed, the momentum fluxes in and out of the control volume cancel out and we have the force balance in terms of shear stress τ,

$$(P|_x - P|_{x+\Delta x})2\pi r\,\Delta r + \tau|_{r+\Delta r}\,2\pi(r+\Delta r)\,\Delta x - \tau|_r\,2\pi r\,\Delta x = 0$$

In terms of the viscosity for a Newtonian fluid, for which $\tau = \mu(\partial u/\partial r)$, the balance may be written

$$(P|_x - P|_{x+\Delta x})2\pi r\,\Delta r + \mu\left.\frac{\partial u}{\partial r}\right|_{r+\Delta r}\,2\pi(r+\Delta r)\,\Delta x - \mu\left.\frac{\partial u}{\partial r}\right|_r\,2\pi r\,\Delta x = 0$$

Dividing through by $2\pi r\,\Delta r\,\Delta x$ and taking the limit as Δx and Δr approach zero, we obtain an equation similar to Eq. (2-2) except that there is an extra term that comes about because of the cylindrical coordinate system.

$$0 = -\frac{\partial P}{\partial x} + \frac{\partial \tau}{\partial r} + \frac{\tau}{r} = -\frac{\partial P}{\partial x} + \frac{1}{r}\frac{\partial(\tau r)}{\partial r} \tag{2-17}$$

and for a Newtonian fluid,

$$0 = -\frac{\partial P}{\partial x} + \mu\frac{\partial^2 u}{\partial r^2} + \frac{\mu}{r}\frac{\partial u}{\partial r} = -\frac{\partial P}{\partial x} + \frac{\mu}{r}\frac{\partial}{\partial r}\left(r\frac{\partial u}{\partial r}\right) \tag{2-18}$$

As before, u is a function only of r and $\partial P/\partial x$ must be a constant. Consequently, we can express the velocity terms as ordinary derivatives,

$$\frac{1}{r}\frac{d}{dr}\left(r\frac{du}{dr}\right) = \frac{1}{\mu}\frac{\partial P}{\partial x} \tag{2-19}$$

and can then integrate with respect to r to obtain

$$\frac{du}{dr} = \frac{r}{2\mu}\frac{\partial P}{\partial x} + \frac{C_1}{r} \tag{2-20}$$

Integrating again, we have

$$u = \frac{r^2}{4\mu}\frac{\partial P}{\partial x} + C_1\ln r + C_2$$

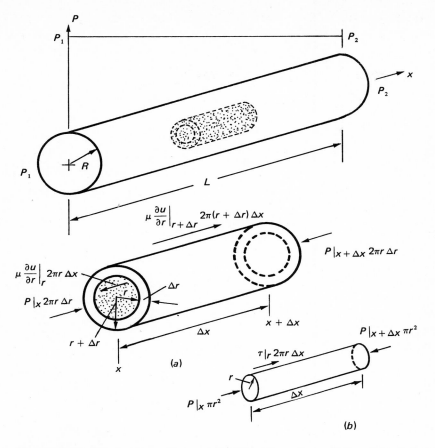

Figure 2-8 (*a*) Circular pipe with a small elemental annular cylinder removed for a force balance. The shear stresses on the surface are written in terms of $\mu \, \partial u/\partial r$ for a Newtonian fluid. If we were to choose a "core" element of radius r (part b) instead of an annular element, the balance would immediately give Eq. (2-20). For flow in circular pipes, such a core-element approach leads more quickly to the final result, but an annular element is a bit more general because it allows for flow calculations in annular flow regions, that is, flow between concentric pipes. Referring to part (*b*), the balance is simply

$$P|_x \, \pi r^2 - P|_{x + \Delta x} \, \pi r^2 + \tau|_r \, 2\pi r \, \Delta x = 0$$

which results in $\partial P/\partial x = 2\tau/r$.

The appropriate boundary conditions are

$$r = 0 \qquad \frac{du}{dr} = 0$$

$$r = R \qquad u = 0$$

From the first condition, C_1 must be zero, and from the second,

$$0 = \frac{R^2}{4\mu} \frac{\partial P}{\partial x} + C_2$$

so that the solution for u is again parabolic in r and the profile is a paraboloid.

$$u = -\frac{1}{4\mu} \frac{\partial P}{\partial x}(R^2 - r^2) \tag{2-21}$$

The flow rate Q is

$$Q = \int_0^R 2\pi r u \, dr = -\frac{\pi R^4}{8\mu} \frac{\partial P}{\partial x} = \frac{\pi R^4}{8\mu L}(P_1 - P_2) = \bar{U}\pi R^2 \tag{2-22}$$

and the mean velocity $\bar{U}$ is $(R^2/8\mu L)(P_1 - P_2)$. If a constant k is defined by $\partial P/\partial x = -k\mu\bar{U}/D^2$, then $k = 32$. Here we have used D instead of h, which was used for the rectangular channel.

The wall shear (in the x direction) in the circular pipe is

$$\tau_w = -\mu \frac{\partial u}{\partial r}\bigg|_{r=R}$$

and is readily found to be

$$\tau_w = -\frac{R}{2} \frac{\partial P}{\partial x} = \frac{R}{2L}(P_1 - P_2) \tag{2-23}$$

An interesting check is to make an overall force balance on the fluid in the pipe. Referring to Fig. 2-9, the total shear force on the fluid must balance the pressure drop across the entire pipe. The shear on the fluid at the wall is $-\tau_w$. Hence,

$$-2\pi R\tau_w L + \pi R^2(P_1 - P_2) = 0 \tag{2-24}$$

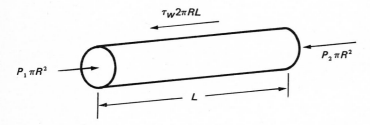

Figure 2-9 An overall force balance on the fluid in a circular pipe.

Substituting for τ_w from above, we see that indeed this equation is satisfied, and thus Eq. (2-24) provides another way to get τ_w without having to solve for the velocity profile.

Another observation we can make at this time is about the relationship between the friction factor and the wall shear we have just calculated. In a pipe the friction factor f is defined in terms of the head loss H_L (in feet of the flowing fluid) or pressure drop in a pipe of diameter D as

$$H_L = \frac{P_1 - P_2}{\rho g} = \frac{\Delta P}{\rho g} = \frac{fL}{D}\frac{\bar{U}^2}{2g} \tag{2-25}$$

From Eq. (2-22) and the definition of the mean velocity, we have

$$f = \frac{64\mu}{\rho D \bar{U}} = \frac{64}{\text{Re}} \tag{2-26}$$

where Re is the dimensionless Reynolds number, $\rho D \bar{U}/\mu$, which we shall discuss later in this chapter. Experimental data of f versus Re are available for frictional flow in pipes, and in the laminar range, where Re is less than about 2300, the above expression is valid.

2-6   FREE SURFACES AND BOUNDARY CONDITIONS AT INTERFACES

The appropriate boundary condition at the wall is the no-slip condition that the velocity of the fluid relative to the wall is zero. At a free surface, that is, a surface of fluid where nothing touches it, the shear must be zero.* Hence we use the condition $\partial u/\partial y = 0$ at a free surface. A liquid-air interface is usually considered a free surface because the shear force of the air on the liquid is negligible.

In general, the boundary condition at any interface between two immiscible fluids is that the velocity is continuous across the interface and the shear is continuous across the interface. That is, in each fluid, just at the edge of the interface, the velocity and shear stress are the same on each side of the interface. However, the derivative $\partial u/\partial y$ is not continuous across the interface unless the viscosity is the same in each fluid. In Fig. 2-10 we choose a thin control volume enclosing a region of the interface. As the control volume shrinks in thickness,

---

*This is not quite correct if surface tension effects are appreciable. Usually they are negligible, but under some conditions, particularly in the flow of liquid metals, they may become important.

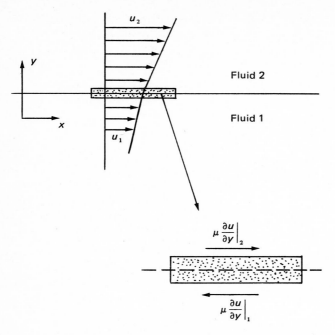

Figure 2-10 A fluid interface showing the continuity of shear stress and velocity. Note that the first derivative $\partial u / \partial y$ is not necessarily continuous but $\mu(\partial u / \partial y)$ and u are.

the momentum flux through the sides approaches zero. From the momentum balance, then, the shears on the opposite flat faces must be equal and opposite. Hence the shear stress must be continuous across the interface and we can state the two conditions

$$\mu_1 \left. \frac{\partial u}{\partial y} \right|_{\text{interface}} = \mu_2 \left. \frac{\partial u}{\partial y} \right|_{\text{interface}}$$

$$u_1 \big|_{\text{interface}} = u_2 \big|_{\text{interface}}$$

(2-27)

as the appropriate boundary conditions on velocity. In three-dimensional flow both components of velocity and shear that lie in the plane of the interface must be continuous.

Let us consider an example of flow with a free surface. In Fig. 2-11 viscous liquid flows down an inclined surface under the action of gravity. We assume the flow to be stable and laminar and the film to be of uniform thickness h. The flow rate Q (per unit width) or film thickness h could be given and the other found. We wish to relate Q and h and find the velocity profile. The pressure forces are zero now, since the free surface is at zero gauge pressure and $\partial P/\partial x = 0$ at the surface. Since we know that $\partial P/\partial x$ is not a function of y, it must be zero throughout the liquid. The gravitational force has a component in the x direction, and the y component generates a hydrostatic pressure that will not concern us here.

Summing forces in the x direction, we have

$$\mu \frac{\partial u}{\partial y}\bigg|_{y+\Delta y} \Delta x - \mu \frac{\partial u}{\partial y}\bigg|_{y} \Delta x + \rho g \sin \theta \, \Delta x \, \Delta y = 0$$

Dividing by $\Delta x \, \Delta y$ and taking the limit, we obtain

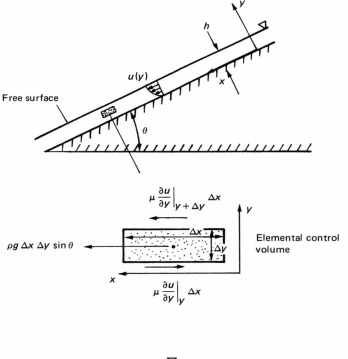

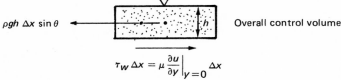

Figure 2-11 Viscous liquid flowing down an inclined surface under the action of gravity. The element shows the forces acting in the x direction.

$$0 = \mu \frac{\partial^2 u}{\partial y^2} + \rho g \sin \theta \qquad (2\text{-}28)$$

which must be solved with the boundary conditions

$$
\begin{aligned}
u &= 0 & y &= 0 \\
\frac{\partial u}{\partial y} &= 0 & y &= h
\end{aligned}
\qquad (2\text{-}29)
$$

Again the equation may be integrated directly, and we find for the velocity profile that

$$u = \frac{\rho g}{\mu} \sin \theta \left(hy - \frac{y^2}{2} \right) \qquad (2\text{-}30)$$

and the flow rate Q in terms of h is

$$Q = \int_0^h u \, dy = \frac{h^3}{3\mu} \rho g \sin \theta \qquad (2\text{-}31)$$

The shear stress on the plane is

$$\tau_w = \mu \left. \frac{\partial u}{\partial y} \right|_{y=0} = \rho g h \sin \theta$$

and can be found directly from Eq. (2-30) or by looking at an overall control volume of fluid as shown in Fig. 2-11. The overall balance gives the result for τ_w directly.

Another example that illustrates a simple case of viscous flow with an interface is the Couette flow of two layers of immiscible liquids. In Fig. 2-12 we show a top plate moving with velocity U, but we assume the pressure gradient to be zero for simplicity. The fluids are of thickness h_1 and h_2, filling the space between the plates. In each fluid we know that the equation of motion is simply

$$0 = \mu \frac{\partial^2 u}{\partial y^2} \qquad (2\text{-}32)$$

since the pressure gradient is zero. The solution is a straight-line profile for $u_1(y)$ and $u_2(y)$.

At the interface the velocity $u_1 = u_2$, which we denote as u_i, and also the shear stress is continuous so that

$$\mu_1 \left. \frac{\partial u_1}{\partial y} \right|_{y=h_1} = \mu_2 \left. \frac{\partial u_2}{\partial y} \right|_{y=h_1}$$

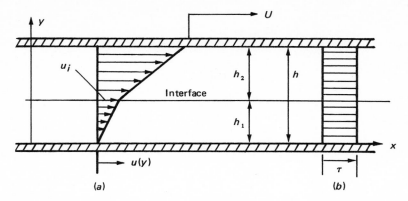

Figure 2-12 Couette flow of two immiscible liquids. (*a*) Velocity profile $u(y)$. (*b*) Shear stress profile. τ is uniform throughout the flow.

which when rewritten in terms of the derivatives is simply

$$\mu_1 \frac{u_i}{h_1} = \mu_2 \frac{U - u_i}{h_2}$$

giving for u_i,

$$u_i = \frac{\mu_2 U h_1}{\mu_1 h_2 + \mu_2 h_1} \tag{2-33}$$

The velocity profiles are simply straight lines as shown in Fig. 2-12, and are completely determined once u_i is found.

2-7 THE CONTINUITY EQUATION IN DIFFERENTIAL FORM

So far we have been using the basic integral forms of the equations of continuity and motion to derive the necessary differential equations for the particular problem we were analyzing. However, it is often convenient to have the necessary differential equations at hand and then to simplify them down to the appropriate form for any particular problem. We shall generally adopt this attitude for the remainder of the book, and only occasionally shall we actually derive a needed differential equation.

There are several ways to determine the differential form of an equation once the integral form is known. We may proceed on a strictly mathematical basis, formally transforming from an integral to a differential form or we may again extract a small elemental volume and apply the integral relationships to a general three-dimensional element.

Let us look at both methods for the continuity equation. In integral form, remember, the equation is for an arbitrary control volume.

$$\frac{\partial}{\partial t} \int_{\text{vol}} \rho \, d(\text{vol}) = - \int_{A} \rho \mathbf{V} \cdot d\mathbf{A} \tag{2-34}$$

where $\mathbf{V}$ is the velocity vector. The term on the left represents the rate of mass storage term and the one on the right the net rate of mass flux into the control volume. Gauss' theorem from vector calculus allows us to transform a surface integral (which completely encloses a volume) into a volume integral throughout that volume. It states that

$$\int_{A} \rho \mathbf{V} \cdot d\mathbf{A} = \int_{\text{vol}} \nabla \cdot (\rho \mathbf{V}) \, d(\text{vol})$$

That is, the integral of the vector flux through the surface area is equal to the integral of the divergence of the integrand throughout the volume. Rearranging the continuity equation, we now have

$$\int_{\text{vol}} \frac{\partial \rho}{\partial t} + \nabla \cdot (\rho \mathbf{V}) \ d(\text{vol}) = 0 \tag{2-35}$$

We have brought the $\partial/\partial t$ operator inside the integral since the volume is constant and the order of differentiation and integration is arbitrary. Equation (2-35) is valid for any arbitrary control volume and hence the only way that this equation can be valid for any volume is if the integrand itself is zero everywhere. By setting the integrand to zero, we immediately have the general equation of continuity in differential form:

$$\frac{\partial \rho}{\partial t} + \nabla \cdot (\rho \mathbf{V}) = 0 \tag{2-36}$$

In Cartesian coordinates, it is

$$\frac{\partial \rho}{\partial t} + \frac{\partial}{\partial x}(\rho u) + \frac{\partial}{\partial y}(\rho v) + \frac{\partial}{\partial z}(\rho w) = 0 \tag{2-37}$$

and the appropriate form in any other coordinate system may be found by expressing the divergence in that coordinate system. Many of these forms are listed in the Appendix of the book. Here u, v, and w are the x, y, and z components of velocity, respectively.

It is rare that the complete form of the equation is needed, and we can often simplify it for special cases. For steady flow, $\partial\rho/\partial t$ is zero, and we have simply

$$\nabla \cdot (\rho\mathbf{V}) = 0 \qquad (2\text{-}38)$$

which may also be written as

$$\mathbf{V} \cdot (\nabla\rho) + \rho\nabla \cdot \mathbf{V} = 0$$

For incompressible flow, which is our concern in this book, the density ρ is constant even for unsteady flow and Eq. (2-36) simplifies to

$$\nabla \cdot \mathbf{V} = 0$$

or, in Cartesian coordinates,

$$\frac{\partial u}{\partial x} + \frac{\partial v}{\partial y} + \frac{\partial w}{\partial z} = 0 \qquad (2\text{-}39)$$

In two dimensions we have only the velocity components u and v, so that

$$\boxed{\frac{\partial u}{\partial x} + \frac{\partial v}{\partial y} = 0} \qquad (2\text{-}40)$$

This is the equation we shall be using extensively later on in two-dimensional boundary-layer theory, and it should be remembered.

In Poiseuille or Couette flow (which is fully developed), we know that $\partial u/\partial x = 0$ and hence $\partial v/\partial y = 0$. As we pointed out earlier, v (the y velocity) was zero there and here we can see that it checks. Since $v = 0$ at the walls, and $\partial v/\partial y = 0$, then v must be zero everywhere in the flow. However, if u were a function of x as in the entrance region, then $\partial u/\partial x$ would not be zero and there would be a y component of velocity, v, in the flow.

For the purposes of illustration it is useful now to apply the integral form of the continuity equation directly to a small element and derive the differential equation that way.

Consider the element in Fig. 2-13 and apply Eq. (2-34). We shall restrict ourselves to Cartesian coordinates, although we could repeat the derivation for any other coordinate system. (The purely mathematical derivation above has the advantage of giving a vector result immediately, which is good for any coordinate system.)

Applying Eq. (2-34), we sum up all the mass fluxes across the surfaces. The product of the velocity normal to a surface, the density at the surface, and the surface area is the mass flow rate across that surface.

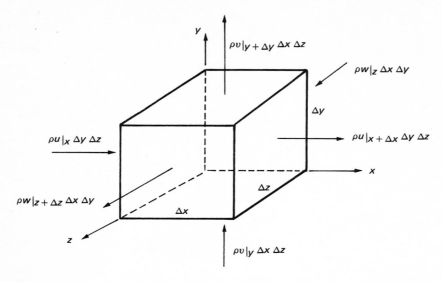

Figure 2-13 An element for the derivation of the equation of continuity in Cartesian coordinates. The fluxes of fluid are shown on each face.

Mass storage rate in the control volume = rate of flow into the control volume
$-$ rate of flow out of the control volume

Or,

$$\frac{\partial}{\partial t}(\rho \, \Delta x \, \Delta y \, \Delta z) = \rho u \big|_x \Delta z \, \Delta y + \rho v \big|_y \Delta x \, \Delta z + \rho w \big|_z \Delta x \, \Delta y$$
$$- \rho u \big|_{x + \Delta x} \Delta z \, \Delta y - \rho v \big|_{y + \Delta y} \Delta x \, \Delta z - \rho w \big|_{z + \Delta z} \Delta x \, \Delta y$$

Dividing by $\Delta x \, \Delta y \, \Delta z$ and taking the limit, we get

$$\frac{\partial \rho}{\partial t} + \frac{\partial}{\partial x}(\rho u) + \frac{\partial}{\partial y}(\rho v) + \frac{\partial}{\partial z}(\rho w) = 0$$

which is the same as we obtained before.

Now let us derive the general equation of motion of a fluid by similar means.

2-8 THE EQUATION OF MOTION IN DIFFERENTIAL FORM

We shall now derive the general equation of motion of a viscous fluid by applying the control volume equation to a small elemental volume. Gauss' law may be applied directly to the integral form, but this becomes rather involved and we shall use the former method instead.

We begin by considering an elemental cube as shown in Figs. 2-14 and 2-15 with the momentum fluxes and stresses shown on the faces. The convention for the stress notation is that the first subscript refers to the face on which the stress acts and the second subscript to the direction of the stress. The face is denoted by the axis to which the face is perpendicular. For example, the stress σ_{xx} is a normal stress acting on the x face in the x direction. Normal stresses are positive for tension and negative for compression. Shear stresses are positive on positive faces if they act in the positive coordinate direction, and on negative faces they are positive if they act in the negative coordinate direction. A positive face is one whose normal points in the positive coordinate direction and a negative face is one whose normal points in the negative coordinate direction.

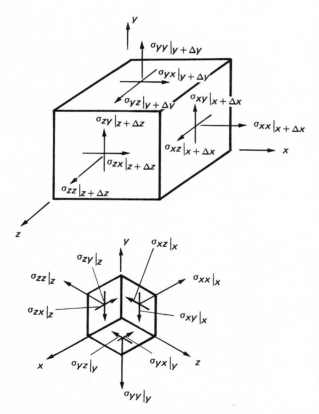

Figure 2-14 Elemental cube for derivation of the equation of motion showing stresses on the positive faces and a rear view showing the stresses on the negative faces.

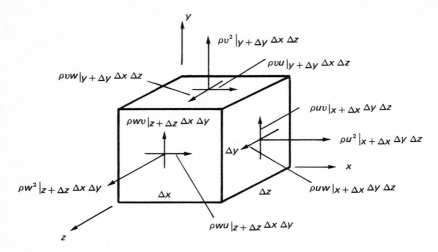

Figure 2-15 The elemental cube of Fig. 2-14 showing momentum fluxes. The positive faces show the momentum fluxes out of the control volume, and on the negative faces the momentum fluxes are oppositely directed and evaluated at x, y, z instead of $x + \Delta x$, $y + \Delta y$, and $z + \Delta z$.

Here we shall derive the equation of motion in terms of the stresses in the fluid. Later we shall relate these stresses to the variables of the fluid flow, the pressure in the fluid, and the viscosity and velocity components.

The integral momentum equation in vector form is

$$\frac{\partial}{\partial t} \int_{\text{vol}} \rho \mathbf{V} \, d(\text{vol}) + \int_A \rho \mathbf{V} \, (\mathbf{V} \cdot d\mathbf{A}) = \mathbf{F}$$

The force $\mathbf{F}$ must include all the stresses on the cube and the left-hand side the momentum fluxes. If a body force, such as that due to gravity or electro-magnetic effects, acts on the fluid, it must also be included in the equation of motion. We shall assume that a vector body force per unit volume $\mathbf{f}$ acts on the fluid. If gravity acts, then the magnitude of $\mathbf{f}$ is ρg.

Referring to Figs. 2-14 and 2-15, we now write out the complete balance. It is convenient to work in Cartesian coordinates and derive one component at a time. Hence we derive only the x component here and from that can infer the form of the y and z components, although the procedure could be repeated for the other components. Similarly, the equation of motion could be derived in other coordinate systems by taking the elemental volume of the appropriate shape. We shall not do that here for the general equation of motion, but the results for other coordinate systems are given in the Appendix. Since our results here will be in Cartesian component form, it would be difficult to infer the general vector form of the equation of motion from it. We shall indicate the general vector form, however, and the form of the equation of motion in other coordinate systems could be derived from the vector form.

The x component of the momentum equation for the element is

$$
\frac{\partial}{\partial t}(\rho u)\,\Delta x\,\Delta y\,\Delta z + \rho u^2\big|_{x+\Delta x}\,\Delta y\,\Delta z - \rho u^2\big|_x\,\Delta y\,\Delta z + \rho v u\big|_{y+\Delta y}\,\Delta x\,\Delta z
$$
$$
- \rho v u\big|_y\,\Delta x\,\Delta z + \rho w u\big|_{z+\Delta z}\,\Delta x\,\Delta y - \rho w u\big|_z\,\Delta x\,\Delta y
$$
$$
= \sigma_{xx}\big|_{x+\Delta x}\,\Delta y\,\Delta z - \sigma_{xx}\big|_x\,\Delta y\,\Delta z + \sigma_{yx}\big|_{y+\Delta y}\,\Delta x\,\Delta z
$$
$$
- \sigma_{yx}\big|_y\,\Delta x\,\Delta z + \sigma_{zx}\big|_{z+\Delta z}\,\Delta x\,\Delta y - \sigma_{zx}\big|_z\,\Delta x\,\Delta y + f_x
$$

Dividing through $\Delta x\,\Delta y\,\Delta z$ and taking the limit as Δx, Δy, and Δz approach zero, we get (remember that the shear stresses must be symmetric, that is, $\sigma_{xy} = \sigma_{yx}$, $\sigma_{xz} = \sigma_{zx}$, and $\sigma_{yz} = \sigma_{zy}$),

$$
\frac{\partial}{\partial t}(\rho u) + \frac{\partial}{\partial x}(\rho u^2) + \frac{\partial}{\partial y}(\rho v u) + \frac{\partial}{\partial z}(\rho w u) = \frac{\partial \sigma_{xx}}{\partial x} + \frac{\partial \sigma_{yx}}{\partial y} + \frac{\partial \sigma_{zx}}{\partial z} + f_x
$$

Combining with the continuity equation (2-37), the terms on the left-hand side simplify and we get a general form good for unsteady compressible flow. We shall write down the y and z components also.

$$
\rho\left(\frac{\partial u}{\partial t} + u\frac{\partial u}{\partial x} + v\frac{\partial u}{\partial y} + w\frac{\partial u}{\partial z}\right) = \frac{\partial \sigma_{xx}}{\partial x} + \frac{\partial \sigma_{yx}}{\partial y} + \frac{\partial \sigma_{zx}}{\partial z} + f_x
$$
$$
\rho\left(\frac{\partial v}{\partial t} + u\frac{\partial v}{\partial x} + v\frac{\partial v}{\partial y} + w\frac{\partial v}{\partial z}\right) = \frac{\partial \sigma_{xy}}{\partial x} + \frac{\partial \sigma_{yy}}{\partial y} + \frac{\partial \sigma_{zy}}{\partial z} + f_y \qquad (2\text{-}41)
$$
$$
\rho\left(\frac{\partial w}{\partial t} + u\frac{\partial w}{\partial x} + v\frac{\partial w}{\partial y} + w\frac{\partial w}{\partial z}\right) = \frac{\partial \sigma_{xz}}{\partial x} + \frac{\partial \sigma_{yz}}{\partial y} + \frac{\partial \sigma_{zz}}{\partial z} + f_z
$$

The terms on the left are acceleration or inertia terms. The time derivative is the unsteady term. The inertia terms involving the velocity gradients are known as convective acceleration terms. Even in steady state, in which quantities at a fixed point in space are constant in time, the fluid can still accelerate of course. As a particle moves along, its velocity may change and give rise to the convective acceleration terms. We are working in Eulerian coordinates, which is a control volume or field theory point of view. If we fix our attention on a particular point in space, we ask what are the properties, as a function of time generally, at that point in space regardless of what particle of fluid is occupying that position in space. Our coordinates refer to position in space and not to the location of any particle of fluid. In particle dynamics, remember, coordinates are usually used to indicate the position of a particle; such a system is known as a Lagrangian coordinate system and is not so useful in fluid mechanics.

Now in a fluid at rest, the mechanical pressure P is isotropic and equal to the negative of the normal stresses, $P = -\sigma_{xx} = -\sigma_{yy} = -\sigma_{zz}$. At this point one is tempted to simplify Eq. (2-41) by the same relationship between pressure and the normal stresses, but this is generally incorrect in a moving fluid. As we shall see presently, the normal stresses are made up of two parts, a compression, indeed due to the isotropic pressure, and an additional part due to viscosity effects when the fluid is in motion. These latter contributions, which may add or subtract from the pressure, are known as deviatoric normal stresses and are not isotropic.

2-9 KINEMATICS OF DEFORMATION AND STRAIN RATES

In the previous section we derived the equation of motion in terms of the stresses acting in a fluid. These results are not very useful as they stand, and eventually we want to relate the stresses to the fluid motion, specifically strain rates, similar to what is done in solid mechanics by Hooke's law.

But first we must discuss the strain rates in a fluid and find them in terms of the velocity components and their spatial derivatives. Let us consider a small element of fluid and ask the question: How can it deform or move? We divide the motion of the element into two types, rigid-body motion and deformation. The center of mass of the element can undergo translation, and the element can undergo rotation. These are the two types of rigid-body motion. In addition, the element can deform as it moves along in translation and rotation. The deformation can be of two types, shear deformation and normal extension or contraction, as shown in Fig. 2-16. The rates of the deformation and the rigid-body motions can be related to the velocity components in the fluid. In reality all these motions may be superimposed to give the net motion of the fluid element.

The three components of the rigid-body translation are merely the velocity components of the velocity vector **V**. In Cartesian coordinates these are u, v, and w in the x, y, and z directions, respectively. For the rotation we must examine a fluid element more closely. In general, the element may rotate with angular velocity Ω in three dimensions. We are interested in determining the deformation and motion for a small element that in the limit becomes infinitesimally small so that these quantities hold at a point in the fluid and represent a state of motion and state of strain rate as a function of position in space.

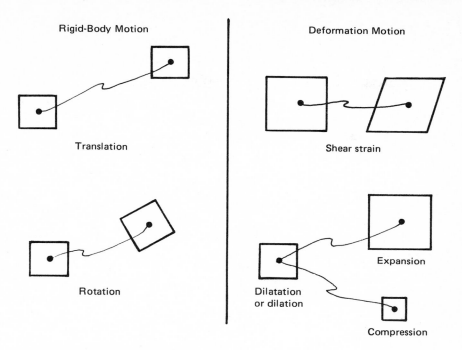

Figure 2-16 The motion and distortion of a fluid element. The two types of rigid-body motion are translation and rotation. The deformation rate consists of shear strain rate and dilation, which is the net expansion or contraction of the fluid element determined by the normal strain rates in the fluid. All these effects may be superimposed to give the net motion of the fluid element.

Rotation

Consider the two-dimensional element, a square Δx by Δy, in Fig. 2-17. Let the element deform only in shear and rotate as shown. The opposite sides of the figure remain parallel. The rotation occurs in the xy plane, yielding the Ω_z component of this rotation vector. The shear strain rate will be denoted as γ_{xy}. On the figure, the rotation of the element is defined as the rotation of the diagonals, which is clearly the average of the angular velocity of the sides of the square. The angular velocity of the side lying on the x axis is $(v|_{x+\Delta x} - v|_x)/\Delta x$, and the angular velocity of the side along the y axis is $-[(u|_{y+\Delta y} - u|_y)/\Delta y]$. We average these two quantities and take the limit as Δx and Δy go to zero to obtain

$$\Omega_z = \frac{1}{2}\left(\frac{\partial v}{\partial x} - \frac{\partial u}{\partial y}\right)$$

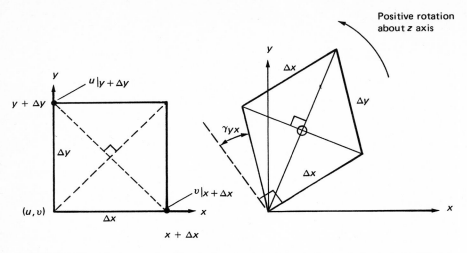

Figure 2-17 A small element in the xy plane undergoing rotation and shear strain. Since we consider no normal strains here, the sides of the square remain fixed in length. We assume unit dimension in the z direction.

In three dimensions the other two components of Ω can be found in a similar manner, and we have, for the components of Ω in Cartesian coordinates,

$$
\begin{aligned}
\Omega_x &= \frac{1}{2}\left(\frac{\partial w}{\partial y} - \frac{\partial v}{\partial z}\right) \\
\Omega_y &= \frac{1}{2}\left(\frac{\partial u}{\partial z} - \frac{\partial w}{\partial x}\right) \\
\Omega_z &= \frac{1}{2}\left(\frac{\partial v}{\partial x} - \frac{\partial u}{\partial y}\right)
\end{aligned}
\tag{2-42}
$$

The concept of vorticity in fluid mechanics is very useful and is used extensively in viscous flow theory. The vorticity vector ω is defined as

$$\omega = 2\Omega \tag{2-43}$$

and in vector notation the vorticity and rotation are

$$
\begin{aligned}
\omega &= \nabla \times \mathbf{V} \\
\Omega &= \tfrac{1}{2}\nabla \times \mathbf{V}
\end{aligned}
\tag{2-44}
$$

which show how they can be found in any coordinate system.

Viscous flows are usually rotational, that is, ω has a finite value throughout the flow. Frictionless flows are generally irrotational, that is, ω is zero throughout the flow unless there are external body forces acting to generate vorticity.

Mathematically it can be shown that if $\nabla \times \mathbf{V} = 0$ (irrotational flow), then the velocity can be derived from a scalar potential Φ as $\mathbf{V} = -\nabla \Phi$. We shall not pursue irrotational flow here, but the subject of potential flow, which we shall mention later in conjunction with boundary-layer theory, forms the foundation of subsonic aerodynamics.

Physically, the vector Ω is the angular velocity of a small element of fluid in the limit as the size becomes infinitesimal. A finite-size element may appear to rotate even in an irrotational flow. A classic example of an irrotational flow is the "potential vortex," which we know by the common example of a whirlpool or tornado. All viscous flows are rotational and have finite rotation throughout. Only flows in which the viscous effects are negligible can approach true irrotational behavior.

One can visualize rotation or vorticity in two-dimensional flow rather easily. Consider a two-dimensional flow with a free surface on which is floated a small cross, made of thin strips of wood, say. In an irrotational flow, such as a whirlpool, the cross will translate in a circular path but will not rotate. In a viscous flow field, such as in Poiseuille flow, the cross will continue to rotate as it moves along.*

Shear Strain Rate

Referring again to Fig. 2-17, the shear strain rate is shown as the rate at which the Δx and Δy baselines rotate toward each other. The Δx baseline rotates with angular velocity $\partial v/\partial x$ toward the y axis, and the Δy baseline rotates toward the x axis with angular velocity $\partial u/\partial y$. Therefore the shear strain rate γ_{yx} must be $(\partial u/\partial y) + (\partial v/\partial x)$ and γ_{xy} is exactly equal to γ_{yx}. Performing the same analysis

---

*A simple experiment may be performed by floating a small piece of wood on a free surface. In particular, the vortex formed by unplugging a bathroom sink provides a useful water surface on which to demonstrate the concept of vorticity.

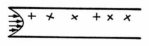

In Poiseuille flow the cross rotates indicating vorticity.

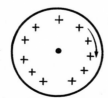

In a vortex the cross remains parallel to itself.

in the other planes, we find generally:

$$\gamma_{xy} = \gamma_{yx} = \frac{\partial u}{\partial y} + \frac{\partial v}{\partial x}$$

$$\gamma_{xz} = \gamma_{zx} = \frac{\partial u}{\partial z} + \frac{\partial w}{\partial x} \qquad (2\text{-}45)$$

$$\gamma_{yz} = \gamma_{zy} = \frac{\partial v}{\partial z} + \frac{\partial w}{\partial y}$$

In any physical situation the numerical values of these strains depend, of course, on the orientation of the axes. Similar equations may be derived for other coordinate systems.

Dilatation

The other type of distortion or deformation that can occur is dilation or contraction of the element. The net rate of such total volume dilation or contraction per unit volume is known as dilatation, written as ϕ, and is the sum of the normal strain rates. (This quantity is numerically independent of the orientation of the axis or coordinate system and is analogous to the simple strain invariant in elasticity.)

Referring to Fig. 2-18, the rate of volume increase for a cube of size Δx by Δy by Δz is clearly

$$(u|_{x+\Delta x} - u|_x)\,\Delta y\,\Delta z + (v|_{y+\Delta y} - v|_y)\,\Delta x\,\Delta z + (w|_{z+\Delta z} - w|_z)\,\Delta x\,\Delta y$$

Dividing through by $\Delta x\,\Delta y\,\Delta z$ and taking the limit, we get the sum of the

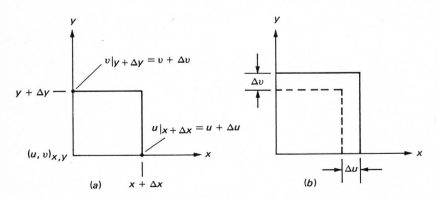

Figure 2-18 An element showing dilatation in two dimensions. (*a*) Initial shape. (*b*) Shape after unit time, referred to the lower left corner as the origin.

normal strains, which is the dilatation:

$$\phi = \frac{\partial u}{\partial x} + \frac{\partial v}{\partial y} + \frac{\partial w}{\partial z} \qquad (2\text{-}46)$$

which in vector form is

$$\phi = \nabla \cdot \mathbf{V} \qquad (2\text{-}47)$$

which is numerically independent of the coordinate system or its orientation. From the continuity equation it can be seen that ϕ is zero for an incompressible fluid.

The Deformation Tensor

Although not essential for what is to follow in this book, it is useful to tie the motion and deformation expressions together in a compact form. By using Cartesian tensor notation the expressions (in Cartesian coordinates) can be greatly compacted and seen in a unified context. We shall assume here that the reader is familiar with Cartesian tensor notation, although if he is not, this section may be either read (since it is fairly self-contained) or skipped entirely.

We write the Cartesian components of the velocity vector $\mathbf{V}$ as u_i, where i takes the numbers 1, 2, or 3 referring to the coordinates x, y, and z, respectively. Thus u_i is another way of writing the components of $\mathbf{V}$. The coordinates are written as x_i where x_1 stands for x, x_2 for y, and x_3 for z. The subscript i is a dummy and may be replaced by j or k, or any other letter for that matter. The derivative $\partial u_i/\partial x_j$ then stands for many terms, all combinations of derivatives of the velocities with respect to the coordinates. A repeated subscript in a single term, such as $\partial u_i/\partial x_i$, means that the subscript i takes on all values 1 through 3, and the individual terms corresponding to each value of i are added. Thus

$$\frac{\partial u_i}{\partial x_i} = \frac{\partial u_1}{\partial x_1} + \frac{\partial u_2}{\partial x_2} + \frac{\partial u_3}{\partial x_3} = \frac{\partial u}{\partial x} + \frac{\partial v}{\partial y} + \frac{\partial w}{\partial z}$$

Now, the term $\partial u_i/\partial x_j$ is known as the *deformation rate tensor* and is really the components of an array of numbers, a matrix, which we can write as

$$\begin{bmatrix} \dfrac{\partial u_1}{\partial x_1} & \dfrac{\partial u_1}{\partial x_2} & \dfrac{\partial u_1}{\partial x_3} \\[2ex] \dfrac{\partial u_2}{\partial x_1} & \dfrac{\partial u_2}{\partial x_2} & \dfrac{\partial u_2}{\partial x_3} \\[2ex] \dfrac{\partial u_3}{\partial x_1} & \dfrac{\partial u_3}{\partial x_2} & \dfrac{\partial u_3}{\partial x_3} \end{bmatrix}$$

Any tensor may be broken up into two parts, a symmetrical and an antisymmetrical part. By formal definition the symmetric part of $\partial u_i/\partial x_j$ is

$$\frac{1}{2}\left(\frac{\partial u_i}{\partial x_j} + \frac{\partial u_j}{\partial x_i}\right)$$

and the antisymmetric part is

$$\frac{1}{2}\left(\frac{\partial u_i}{\partial x_j} - \frac{\partial u_j}{\partial x_i}\right)$$

Added together they give simply $\partial u_i/\partial x_j$. These two parts are related to the components of the rotation and shear strain rate. Referring to Eq. (2-45), we see that the shear strain rates are

$$\gamma_{ij} = \frac{\partial u_i}{\partial x_j} + \frac{\partial u_j}{\partial x_i} \tag{2-48}$$

and we can define a new quantity, a symmetrical tensor, called the shear strain rate tensor e_{ij}:

$$e_{ji} = e_{ij} = \frac{1}{2}\gamma_{ij} = \frac{1}{2}\left(\frac{\partial u_i}{\partial x_j} + \frac{\partial u_j}{\partial x_i}\right) \tag{2-49}$$

which is just the symmetrical part of the deformation rate tensor and may be compared with Eq. (2-45). The components of the strain rate tensor may be put in matrix form as follows, where the diagonal terms are the normal strain rates and the off-diagonal terms (which are symmetric) are equal to one-half the shear strain rates:

$$\begin{bmatrix} e_{11} & e_{12} & e_{13} \\ e_{21} & e_{22} & e_{23} \\ e_{31} & e_{32} & e_{33} \end{bmatrix} \tag{2-50}$$

The diagonal terms are exactly the normal strain rates, for example, $e_{11} = e_{xx} = \partial u_1/\partial x_1$, but the shear terms of the tensor are one-half of the true shear strains. The reason for introducing the factor $\frac{1}{2}$ is to make e_{ij} a true tensor so that the normal or diagonal terms and the shear terms can be written simply as indicated in Eq. (2-49).

The sum of the diagonal terms, the trace of the matrix (2-50), is the dilatation,

$$\phi = e_{ii} = e_{11} + e_{22} + e_{33} \tag{2-51}$$

and we know from formal mathematics that the trace of a matrix is an invariant independent of the orientation of the coordinate system. The value of ϕ is zero for an incompressible fluid.

Similarly, we can define an antisymmetric tensor Ω_{ij} (such that $\Omega_{ij} = -\Omega_{ji}$) as

$$\Omega_{ij} = -\Omega_{ji} = \frac{1}{2}\left(\frac{\partial u_i}{\partial x_j} - \frac{\partial u_j}{\partial x_i}\right) \tag{2-52}$$

Comparing this expression to Eqs. (2-42) and (2-43), we see that the components of Ω and Ω_{ij} are related as

$$\Omega_1 = \frac{1}{2}\left(\frac{\partial u_3}{\partial x_2} - \frac{\partial u_2}{\partial x_3}\right) = \Omega_{32} = -\Omega_{23}$$

$$\Omega_2 = \frac{1}{2}\left(\frac{\partial u_1}{\partial x_3} - \frac{\partial u_3}{\partial x_1}\right) = \Omega_{13} = -\Omega_{31} \tag{2-53}$$

$$\Omega_3 = \frac{1}{2}\left(\frac{\partial u_2}{\partial x_1} - \frac{\partial u_1}{\partial x_2}\right) = \Omega_{21} = -\Omega_{12}$$

(Mathematically, Ω is an axial vector and it can be shown that to every axial vector there corresponds an antisymmetric tensor.) We have used the notation Ω_i for the rotation vector and Ω_{ij} for the components of the rotation tensor. The use of the same symbol here for both should not cause any confusion.

Hence we can write the deformation tensor as

$$\frac{\partial u_i}{\partial x_j} = e_{ij} + \Omega_{ij} \tag{2-54}$$

Next, we shall relate the stresses in a fluid to these deformation quantities we have been discussing. The stresses will be related to the deformation quantities e_{ij} and ϕ, but will be independent of the rigid-body motions, u_i and Ω_{ij}.

2-10 STRESS–STRAIN RELATIONSHIPS AND THE NAVIER–STOKES EQUATION

We have mentioned that the state of stress in a fluid is given by the stress components σ_{ij}, where i and j can take on the values of the coordinates. In Cartesian coordinates the Cartesian tensor σ_{ij} represents the components of the Cartesian stress tensor. Remember, the first subscript refers to the face on which the stress acts and the second subscript is the direction in which the stress acts. The components may be written in matrix form similar to Eq. (2-50). The stress and strain rate tensors may be related by phenomenological equations, which allow the equation of motion (2-41) to be expressed in terms of the velocity components instead of stresses that cannot be measured directly.

For many fluids, the stress tensor σ_{ij} and the strain rate tensor e_{ij} may be related by a linear function to a high degree of accuracy. In the introduction we mentioned a Newtonian fluid, and now we can give a more precise definition of a Newtonian fluid as one for which this linear relationship holds. Most common fluids, such as water, oil, and gases, are Newtonian fluids. However, there are some very important classes of fluids that are not Newtonian and require nonlinear, sometimes complex, relationships between stress and strain rate. We shall discuss some of these fluids in the next section.

We shall not derive the most general form of the linear stress-strain rate relationships (since it is rather lengthy), but shall merely list it here for reference and point out the assumptions that underlie its derivation.

The basic assumptions are as follows:

1. The stress tensor σ_{ij} is symmetric, that is, $\sigma_{ij} = \sigma_{ji}$. This can be seen from equilibrium considerations. If we look at an element and write a dynamical equation for its rotational motion (with respect to an observer fixed with respect to its center of mass), we have, referring to Fig. 2-19, considering an element Δx by Δy by unit thickness in the z direction,

$$(\sigma_{xy}|_{x+\Delta x} + \sigma_{xy}|_x)\frac{\Delta x}{2} - (\sigma_{yx}|_{y+\Delta y} + \sigma_{yx}|_y)\frac{\Delta y}{2} = I_z \dot{\Omega}_z \qquad (2\text{-}55)$$

I_z, the polar moment of inertia of the element, is proportional to $(\Delta x\, \Delta y)^2$, so that as Δx and Δy approach zero, the left-hand side of Eq. (2-55) must be zero in order for Ω_z to remain finite. Hence we conclude that $\sigma_{xy} = \sigma_{yx}$. A

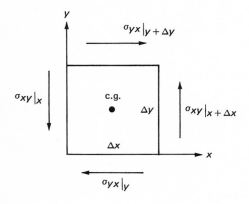

Figure 2-19 Dynamics of a rotating element acted upon by shear stresses.

similar procedure may be carried out for the other components of Ω to show that $\sigma_{ij} = \sigma_{ji}$.

2. The fluid pressure P may be separated out of the stress tensor by subtracting it from the diagonal terms. In a fluid at rest, the fluid pressure P is exactly equal to the negative of each of the diagonal terms. However, in moving viscous fluids there may be some contributions to the normal stress from viscous effects. We can separate σ_{ij} into a part due to pressure P and a part due entirely to motion σ'_{ij}, known as the deviatoric stress tensor.

$$\sigma_{ij} = \sigma'_{ij} - P\delta_{ij} \tag{2-56}$$

Here δ_{ij} is the Kronecker delta, defined as

$$\begin{aligned} \delta_{ij} &= 1 \quad \text{for} \quad i = j \\ \delta_{ij} &= 0 \quad \text{for} \quad i \neq j \end{aligned} \tag{2-57}$$

The mechanical pressure P is defined as $P = -\frac{1}{3}(\sigma_{11} + \sigma_{22} + \sigma_{33}) = -\frac{1}{3}\sigma_{ii}$ and is an invariant of the stress tensor.* The normal deviatoric stresses σ'_{11}, σ'_{22}, σ'_{33} are then the differences between the actual normal stresses σ_{11}, σ_{22}, σ_{33} and their average. In effect the normal components of σ'_{ij} reflect the anisotropy of the normal stresses.

In terms of these defined quantities now we can write the general linear stress-strain rate relationships for a Newtonian fluid:

$$\sigma_{ij} = \sigma'_{ij} - P\delta_{ij} = -P\delta_{ij} + 2\mu e_{ij} + \delta_{ij}\lambda\phi \tag{2-58}$$

Here two new symbols, μ and λ, have been introduced. The first, μ, is the fluid viscosity, which we have discussed in the introduction, and the second, λ, is called the second coefficient of viscosity and is sometimes important in compressible flow, particularly in acoustics and wave motion. For reference we mention that another definition of the second coefficient of viscosity, ζ, is in current usage and is given by

$$\zeta = \lambda + \tfrac{2}{3}\mu \tag{2-59}$$

The general vector form of Eq. (2-58) will not be given here, but the explicit form of Eq. (2-58) may be written out (where $\phi = \partial u/\partial x + \partial v/\partial y + \partial w/\partial z$) in

---

*The mechanical pressure P defined here cannot be identified directly with the thermodynamic pressure except in certain special cases. In a gas, for example, thermodynamic equilibrium must be established, which implies negligible dissipation. A general discussion is beyond our needs in this book.

Cartesian coordinates as

$$\sigma_{xx} = -P + 2\mu \frac{\partial u}{\partial x} + \lambda \phi$$

$$\sigma_{yy} = -P + 2\mu \frac{\partial v}{\partial y} + \lambda \phi$$

$$\sigma_{zz} = -P + 2\mu \frac{\partial w}{\partial z} + \lambda \phi$$

$$\sigma_{xy} = \sigma_{yx} = 2\mu e_{xy} = \mu\left(\frac{\partial u}{\partial y} + \frac{\partial v}{\partial x}\right)$$

$$\sigma_{xz} = \sigma_{zx} = 2\mu e_{xz} = \mu\left(\frac{\partial w}{\partial x} + \frac{\partial u}{\partial z}\right)$$

$$\sigma_{yz} = \sigma_{zy} = 2\mu e_{yz} = \mu\left(\frac{\partial v}{\partial z} + \frac{\partial w}{\partial y}\right)$$

(2-60)

By substituting the expressions (2-60) into the equations of motion (2-41), we finally obtain the working equations of viscous flow known as the Navier-Stokes equations. For completeness we write down this equation in vector form:

$$\rho\left[\frac{\partial \mathbf{V}}{\partial t} + \nabla\left(\frac{V^2}{2}\right) - \mathbf{V} \times (\nabla \times \mathbf{V})\right] = -\nabla P + \mathbf{f} - \nabla \times [\mu(\nabla \times \mathbf{V})] + \nabla[(\lambda + 2\mu)\nabla \cdot \mathbf{V}]$$

(2-61)

But we are not really interested in such complications here, and we now write out the equations of immediate interest to us, the equations of motion in Cartesian coordinates for a viscous incompressible fluid with constant viscosity:

$$\rho\left(\frac{\partial u}{\partial t} + u\frac{\partial u}{\partial x} + v\frac{\partial u}{\partial y} + w\frac{\partial u}{\partial z}\right) = -\frac{\partial P}{\partial x} + f_x + \mu\left(\frac{\partial^2 u}{\partial x^2} + \frac{\partial^2 u}{\partial y^2} + \frac{\partial^2 u}{\partial z^2}\right)$$

$$\rho\left(\frac{\partial v}{\partial t} + u\frac{\partial v}{\partial x} + v\frac{\partial v}{\partial y} + w\frac{\partial v}{\partial z}\right) = -\frac{\partial P}{\partial y} + f_y + \mu\left(\frac{\partial^2 v}{\partial x^2} + \frac{\partial^2 v}{\partial y^2} + \frac{\partial^2 v}{\partial z^2}\right)$$

$$\rho\left(\frac{\partial w}{\partial t} + u\frac{\partial w}{\partial x} + v\frac{\partial w}{\partial y} + w\frac{\partial w}{\partial z}\right) = -\frac{\partial P}{\partial z} + f_z + \mu\left(\frac{\partial^2 w}{\partial x^2} + \frac{\partial^2 w}{\partial y^2} + \frac{\partial^2 w}{\partial z^2}\right)$$

(2-62)

These represent a formidable set of equations indeed, and they will have to be cut down to size before we can do anything very useful with them. For example, the first equation with the left-hand side zero, no body force, and no variations in the x or z direction is precisely Eq. (2-2).

There are two approaches, then, to finding an appropriate differential equation describing a physical model of a flow situation. We may either start from an elemental volume and derive an appropriate equation under our assumptions and consistent with our physical model, or we may begin directly with the equation of motion (2-61) or (2-62) (if the flow is incompressible) and whittle the equation down to size by neglecting terms that are small under our assumptions.

Both methods require a certain skill that comes only from problem-solving experience, and both should be understood by the serious student who wants to be able to apply these rather formal mathematical relationships to real engineering problems.*

2-11 THE REYNOLDS NUMBER AND NORMALIZATION OF THE EQUATION OF MOTION

We have mentioned the Reynolds number before, but now we can show how it appears as an important parameter by normalizing (or nondimensionalizing) the

---

*The Navier-Stokes equation may be written in terms of vorticity ω. Such a formulation is extremely useful for many general two- and three-dimensional viscous flow problems. Although we shall not pursue such problems in this book, the vorticity form of the equation of motion is presented here for reference.

Consider the equation of motion for a viscous incompressible fluid with constant viscosity. Equation (2-61) takes the form

$$\rho\left[\frac{\partial \mathbf{V}}{\partial t} + \nabla\left(\frac{V^2}{2}\right) - \mathbf{V} \times (\nabla \times \mathbf{V})\right] = -\nabla P + \mathbf{f} - \mu\nabla \times (\nabla \times \mathbf{V})$$

where the body force $\mathbf{f}$ is assumed to be derived from a conservative force field so that it can be written as the gradient of a scalar potential (see Section 2-13). By taking the curl of the above equation and remembering that the curl of the gradient of a scalar is zero, we obtain

$$\frac{\partial \omega}{\partial t} - \nabla \times (\mathbf{V} \times \omega) = \nu\nabla^2 \omega$$

where ν is the kinetic viscosity μ/ρ. This equation is known as the vorticity transport equation. It is a convenient form of the equation of motion since the pressure drops out and allows a dynamical description in terms of vorticity alone. Of course the continuity equation must be satisfied along with the vorticity transport equation. For two-dimensional flow a stream function Ψ may be introduced, which ensures satisfaction of continuity, as

$$u = -\frac{\partial \Psi}{\partial y} \qquad v = \frac{\partial \Psi}{\partial x}$$

and the vorticity equation written in terms of the stream function becomes a scalar equation instead of a vector equation and is used extensively for two-dimensional flow analysis:

$$\frac{\partial}{\partial t}(\nabla^2 \Psi) - \frac{\partial \Psi}{\partial y}\frac{\partial}{\partial x}(\nabla^2 \Psi) + \frac{\partial \Psi}{\partial x}\frac{\partial}{\partial y}(\nabla^2 \Psi) + \nu\nabla^4 \Psi$$

A discussion of the solution is beyond the scope of this text.

equation of motion. If any basic equation of fluid mechanics is nondimensionalized, parameters appear as coefficients of certain of the terms in the equation. The parameters are dimensionless and their numerical values characterize the fluid and its flow.

For example, let us consider the equation of motion. Generally, we could begin with the vector equation of motion, but for our purposes it is adequate to examine one component, say, the x component in two-dimensional flow. Assuming an incompressible fluid with no body force and in steady state, we have

$$\rho\left(u\frac{\partial u}{\partial x} + v\frac{\partial u}{\partial y}\right) = -\frac{\partial P}{\partial x} + \mu\left(\frac{\partial^2 u}{\partial x^2} + \frac{\partial^2 u}{\partial y^2}\right) \tag{2-63}$$

Let us introduce the dimensionless variables denoted by primes as

$$x' = \frac{x}{L}$$

$$y' = \frac{y}{h}$$

$$u' = \frac{u}{U} \tag{2-64}$$

$$v' = \frac{v}{U}\frac{L}{h}$$

$$P' = \frac{P}{P_0}$$

where L and h are characteristic lengths of the flow in the x and y directions, respectively, U is a characteristic velocity in the x direction, and hU/L is taken as the characteristic velocity in the y direction. P_0 is a characteristic pressure. We now replace the dimensional variables in Eq. (2-63) by their values in terms of the dimensionless variables and characteristic numbers. Rearranging a bit, we readily obtain

$$u'\frac{\partial u'}{\partial x'} + v'\frac{\partial v'}{\partial y'} = -\frac{P_0}{\rho U^2}\frac{\partial P'}{\partial x'} + \frac{1}{\mathrm{Re}_L}\left(\frac{\partial^2 u'}{\partial x'^2} + \frac{L^2}{h^2}\frac{\partial^2 u'}{\partial y'^2}\right) \tag{2-65}$$

where $\mathrm{Re}_L = \rho UL/\mu$ and is the Reynolds number based on the characteristic length in the x direction. The term $P_0/\rho U^2$ is a dimensionless pressure coefficient.

We see that since all the variables are now dimensionless, their order of magnitude is unity. Thus the ratio of the inertia term (the left-hand side) to the viscous terms is the Reynolds number, Re_L or $Re_L(h^2/L^2)$. Hence the Reynolds number is a measure of the importance of inertia effects compared to viscosity effects in a flowing fluid. If Re_L is small compared to unity, and $h \approx L$, then the viscous terms dominate the inertia terms. However, we must examine both Re_L and $Re_L(h^2/L^2)$.* If, as is the case in many quasi-one-dimensional flows where there are essentially no velocity variations in the x direction, the criterion for negligible inertia is that $Re_L(h^2/L^2)$ be much smaller than unity. Often in physical situations $h \ll L$, as is the case in lubrication-type flow, where $h \approx 10^{-3}L$, and then the value of Re_L itself may be larger than unity and inertia effects still negligible.

The same analysis may be extended to three dimensions with the same results except that generally we need consider Re_L, h/L, and another ratio l/L, where l is a characteristic dimension in the z direction. Then, for the y and z equation of motion, Re will be based on h and l instead of L.

Many other dimensionless parameters arise in fluid mechanics. In the equation of motion, if we had retained a gravitational body force, the Froude number, U^2/gL, would have appeared. This number is a measure of the ratio of inertia to gravitational force. In the energy equation other important parameters may be formed, and we shall derive some of them later when we discuss heat transfer and thermal effects in viscous flow, and in particular boundary-layer flow.

The formulation of these dimensionless parameters in fluid mechanics forms the foundation of experimental modeling theory. Consider a small model of a fluid flow that is geometrically similar to a large prototype. The descriptive differential equations and boundary conditions in dimensionless parameters will be the same for model and prototype if the dimensionless parameters, such as the Reynolds number, pressure coefficient, and so on, are the same for model and prototype. Hence the value of dimensionless variables measured on the model will be identical to those expected on the prototype. In wind tunnel testing of aerodynamic objects, the model is made geometrically similar to the prototype and the relevant parameters, say Reynolds number and/or Mach number, are made the same as expected for the prototype in real performance. Dimensionless variables and performance characteristics, such as lift and drag coefficients, can then be measured on the model and scaled for the real prototype.

Later on we shall be particularly interested in some of the parameters of use in heat flow in boundary layers.

---

*The quantity $Re_L(h^2/L^2)$ is sometimes called the reduced Reynolds number and is of importance in lubrication theory, as we shall see in the next chapter.

2-12 NON–NEWTONIAN FLUIDS

Although many fluids in nature are Newtonian (that is, the stress and strain rate tensors are linearly related), there are many fluids of practical importance that do not even approximate a linear relationship. Grease, paint, ink, slurries, and some polymers are just a few examples of non-Newtonian fluids.

There are three general classifications of non-Newtonian fluids:

1. Time-independent non-Newtonian fluids, in which the shear strain rate is a single-valued but nonlinear function of shear stress.
2. Time-dependent non-Newtonian fluids, in which the shearing rate is not a single-valued function of the shear stress and may depend on the previous shear stress history of the fluid. The shear strain rate versus shear stress curve may form a hysteresis loop.
3. Viscoelastic fluids, in which shear strain as well as shear strain rate are related to the shear stress. Because of the partially elastic property of these fluids, not all energy of shear deformation is dissipated (as in viscous fluids) but some may be stored in the fluid as elastic strain energy. In time-varying processes the stress-strain rate relationship may be a function of frequency, which adds to the complication of mathematically modeling a viscoelastic material.

We shall look at these non-Newtonian fluids in more detail and show how the shear stress and strain rate may be related by a mathematical model.

Time-Independent Fluids

In general, the strain rate term e_{ij} may be related to the stress tensor σ_{ij} as an arbitrary function

$$\sigma_{ij} = f(e_{ij}) \tag{2-66}$$

which for Newtonian fluids is the linear relationship (2-58). It is very difficult to devise appropriate nonlinear relationships for general three-dimensional flow, and most simple models of non-Newtonian flow are developed on the basis of one- and two-dimensional flow. If we consider a flow in which there is only one component of true shear strain rate, which we denote as γ, and one component of shear stress, which we write as τ, we can illustrate the various classes of time-independent non-Newtonian flow on a γ versus τ curve, Fig. 2-20.

There are three distinct types of fluids shown in Figure 2-20: the Bingham plastic A, pseudoplastic fluid B, and dilatant fluid C, in addition to the Newtonian fluid D.

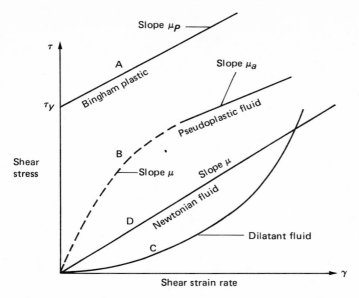

Figure 2-20 Example of stress-strain rate relationship for time-independent non-Newtonian fluids.

A Bingham plastic behaves like a solid until the yield stress τ_y is exceeded; then it behaves like a viscous Newtonian fluid with viscosity μ_P. The shear stress may be expressed as

$$\tau = \tau_y + \mu_P \gamma \quad \tau > \tau_y$$
$$\gamma = 0 \qquad \tau < \tau_y \tag{2-67}$$

The Bingham plastic concept is a close approximation to many real fluids: some slurries, greases, paint, and some suspensions of finely divided particles such as clay in water.

Pseudoplastic fluids and dilatant fluids do not have a yield stress. The pseudoplastic fluid has a progressively decreasing slope of shear stress versus shear strain rate (apparent viscosity), which tends to become constant for very large values of shear stress. Many models have been developed to describe pseudoplastic flow, but the simplest and perhaps most useful is the power-law relationships due to Ostwald,

$$\tau = k\gamma^n \quad (n < 1) \tag{2-68}$$

where k and n are constants for a particular fluid. The parameter k is a measure of the "consistency" of the fluid, and n is a measure of how much the fluid deviates from a Newtonian fluid for which $k = \mu$ and $n = 1$. Several more complex models of pseudoplastic fluids have been proposed, but they lead to rather complex representations and will not be discussed here.

Dilatant fluids are similar to pseudoplastic fluids except that the apparent viscosity increases with strain rate. Power-law models may also be used here, except that the exponent n is now greater than unity.

Time-Dependent Fluids

Time-dependent fluids have an apparent viscosity that depends not only on the shear rate, as just described above, but also on the time for which the shear has been applied. There are two general classes of time-dependent fluids, thixotropic fluids and rheopectic fluids. The shear stress decreases with time in a thixotropic fluid and increases with time in a rheopectic fluid. Printer's ink is an example of a thixotropic fluid, and egg white behaves like a rheopectic fluid (although it tends to exhibit some viscoelastic properties as it becomes stiff and is not a true rheopectic fluid). Many fluids lose their rheopectic property at very high shear rates and may then behave as thixotropic fluids. An experienced homemaker knows that many food preparations must be beaten slowly and for a limited time or they will cease to stiffen and become watery again.

The behavior of a thixotropic fluid is illustrated by Fig. 2-21. The apparent viscosity depends on both the shear rate and the length of time of shearing. If

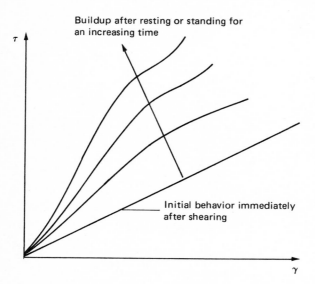

Buildup after resting or standing for an increasing time

Initial behavior immediately after shearing

Figure 2-21 The behavior of a thixotropic fluid allowed to rest for differing times from a state of just having been sheared.

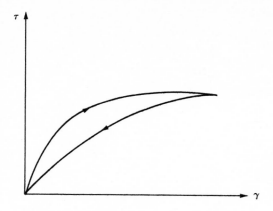

Figure 2-22 Hysteresis loop found by shearing a thixotropic fluid at a constantly increasing rate and then at a constantly decreasing rate.

the fluid is sheared from a state of rest, it breaks down and the apparent viscosity decreases. This breakdown actually occurs on a molecular scale. If the fluid is allowed to rest for a certain time, the consistency builds up; and if it is allowed to rest for a long enough time, the fluid regains its original consistency. In Fig. 2-21 the initial curve represents the behavior immediately after the fluid has been rigorously sheared or beaten and is shown as a Newtonian curve although it could just as well be a pseudoplastic-type curve.

Thixotropic fluids exhibit hysteresis behavior. If they are sheared at a constantly increasing rate, then at a constantly decreasing rate, a closed hysteresis loop is formed as shown in Fig. 2-22.

Viscoelastic Fluids

Viscoelastic materials can be rather complex and exhibit both viscous and elastic-type behavior. The simplest model of a viscoelastic material is the Maxwell liquid model, which assumes that the fluid is Newtonian in viscosity and obeys Hooke's law for the elastic behavior. The stress-strain rate relationship is

$$\tau + \left(\frac{\mu}{\lambda}\right)\dot{\tau} = \mu\gamma \tag{2-69}$$

where $\dot{\tau}$ is the time derivative of stress state, μ is the viscosity, and λ is the modulus of rigidity. If the flow is steady, $\tau = \mu\gamma$ and the material behaves like a Newtonian fluid, but under time-varying stress, an elastic effect enters into the analysis. Physically, the constant (λ/μ) is a relaxation time. If motion is stopped $(\gamma = 0)$, the stress decays (at constant strain) as $\exp(-\lambda t/\mu)$ and λ/μ is the time constant for that decay.

Laminar Flow of a Power–Law Fluid in a Circular Pipe

As an example of non-Newtonian flow, we can solve for the velocity profiles of laminar flow in a pipe. The power-law fluid and Bingham plastic are two types of fluids that yield simple solutions for pipe flow.

First consider the power-law fluid. We could begin with the equation of motion in terms of stress, Eq. (2-41) (but in cylindrical coordinates), and replace the relevant stress term with the appropriate stress-strain rate relationship; or we can examine a small element of fluid and derive the necessary differential equation. By considering an element as shown in Fig. 2-23, we can obtain a differential equation in terms of τ immediately. (An annular element as shown in Fig. 2-8 would also be appropriate, but since it would give a derivative of the shear stress τ and a cylindrical elemental control volume gives τ directly, we can save one integration this way.) From Fig. 2-23 and a force balance, we have

$$(P|_{x+\Delta x} - P|_x)\pi r^2 - \tau 2\pi r\,\Delta x = 0$$

and dividing by Δx and taking the limit we have

$$\tau = \frac{r}{2}\left(\frac{\partial P}{\partial x}\right) \tag{2-70}$$

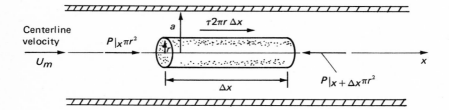

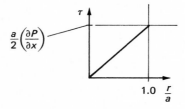

Figure 2-23 Cylindrical element for a force balance. By equating the shear stress to pressure gradient we obtain a differential equation in terms of stress. The annular element of Fig. 2-8 could also be used to find the equation, but the cylindrical element gives a first integral, that is, it involves the stress directly and not its derivative.

It is interesting to note that this equation could have been obtained directly from Eq. (2-17) by a single integration. Further, the development here is identical to that indicated in Fig. 2-8. For positive flow (negative $\partial P/\partial x$), τ will be a negative number. Following convention, we have shown τ positive in the positive x direction on the positive r face. Again, $\partial P/\partial x$ must be a constant in fully developed flow, since the velocity u and the shear stress τ are functions of r only. If any hydrostatic variation in the r direction is neglected, then $\partial P/\partial x$ may be written as dP/dx, which is customary in engineering work. In what follows we shall denote the negative pressure gradient $(-dP/dx)$ as G. Then a positive value of G gives rise to positive flow (in the x direction).

For a power-law fluid, τ is related to the velocity gradient in a circular pipe as

$$\tau = -k\left(-\frac{du}{dr}\right)^n \tag{2-71}$$

for values of du/dr negative (i.e., flow in the positive x direction). The sign of the quantity in parentheses in Eq. (2-71) must be positive, and the correct sign for τ must be obtained by the correct sign outside the parentheses.

Equation (2-71) is a bit awkward and, in general for a power-law fluid (with a one-dimensional velocity profile), we may express τ as

$$\tau = k\epsilon \left|\frac{du}{dy}\right|^n$$

where ϵ is a sign factor given by

$$\epsilon = \frac{du/dy}{|du/dy|}$$

so that

$$\tau = k\left|\frac{du}{dy}\right|^{n-1}\frac{du}{dy} \tag{2-72}$$

We see that $k|du/dy|^{n-1}$ plays the role of a local apparent viscosity.

Equation (2-70) can be written explicitly (if we assume that du/dr is negative, corresponding to flow in the positive x direction)

$$k\left(-\frac{du}{dr}\right)^n = \frac{r}{2}G \tag{2-73}$$

Solving for du/dr,

$$\frac{du}{dr} = -\left(\frac{G}{2k}\right)^{1/n}r^{1/n} \tag{2-74}$$

We can integrate directly between limits,

$$\int_u^0 -du = \int_r^a \left(\frac{G}{2k}\right)^{1/n} r^{1/n} \, dr$$

to give the velocity profile

$$u = \frac{n}{n+1}\left(\frac{G}{2k}\right)^{1/n}(a^{(n+1)/n} - r^{(n+1)/n}) \tag{2-75}$$

so that the flow rate is

$$Q = \int_0^a 2\pi u r \, dr = \frac{n\pi}{3n+1}\left(\frac{G}{2k}\right)^{1/n} a^{(3n+1)/n} \tag{2-76}$$

and the mean velocity is

$$\bar{U} = \frac{Q}{\pi a^2} = \frac{n}{3n+1}\left(\frac{G}{2k}\right)^{1/n} a^{(n+1)/n} \tag{2-77}$$

Plots of the velocity profile for various values of n are shown in Fig. 2-24. The profiles flatten out as n decreases; and as n approaches zero, the flow becomes

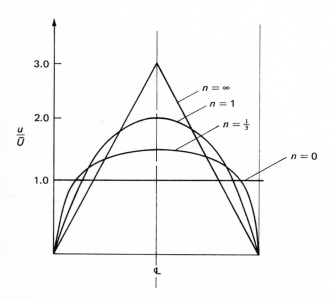

Figure 2-24 Velocity profiles for a power-law fluid flow in a circular pipe.

pluglike with a shear layer next to the walls similar to the Bingham plastic profile. At $n = 1$ the flow is Newtonian.

Laminar Flow of a Bingham Plastic in a Circular Pipe

For laminar flow of a Bingham plastic, Eq. (2-70) is still valid. But now instead of Eq. (2-71) we have the relationship (2-67),

$$\tau = \tau_y + \mu_P \gamma$$

in the flow region where $\tau > \tau_y$. Where $\tau < \tau_y$, the shear strain rate γ is zero and no relative motion occurs. Hence the fluid must flow as a plug of radius r_P in the center of the pipe, where $\tau < \tau_y$, and in a Newtonian-like profile in the region near the wall, where $\tau > \tau_y$ (Fig. 2-25). The shear stress varies across the pipe as indicated in Fig. 2-23 and by Eq. (2-70); that is, it varies from zero at the pipe center linearly to a maximum value at the wall. Clearly, there is a critical value of G below which no flow will occur. This value is obtained by setting $\tau = \tau_y$ at $r = a$ in Eq. (2-70); $G_{cr} = 2\tau_y/a$.

In the shear region, $r > r_P$, the equation of motion obtained by substituting (2-67) into (2-70) is

$$\tau_y - \mu_P \frac{du}{dr} = \left(\frac{r}{2}\right) G \tag{2-78}$$

for $r_P < r < a$. But for $r < r_P$, $du/dr = 0$. At $r = r_P$ the shear is τ_y and, since the shear is continuous, $du/dr = 0$ at r_P and is continuous.

We integrate Eq. (2-78) for the velocity using the boundary conditions

$$u = 0 \qquad r = a$$

$$\frac{\partial u}{\partial r} = 0 \qquad u = u_P \qquad r = r_P \tag{2-79}$$

to find the one constant of integration and the value of r_P and u_P. We obtain

$$u = \frac{G}{4\mu_P}(a^2 - r^2) - \frac{\tau_y}{\mu_P}(a - r) \qquad r_P < r < a \tag{2-80}$$

$$r_P = \frac{2\tau_y}{G} \tag{2-81}$$

$$u_P = \frac{\tau_y^2}{\mu_P G}\left(\frac{a}{r_P} - 1\right)^2 \qquad r_P > r > 0 \tag{2-82}$$

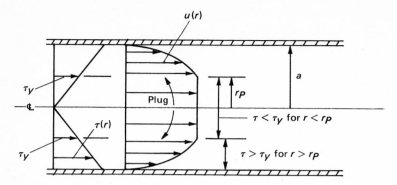

Figure 2-25 Bingham plastic flow in a circular pipe. Plug flow occurs for $r < r_P$.

The flow Q is then

$$Q = \int_{r_P}^{a} u 2\pi r \, dr + \pi r_P^2 u_P = \frac{\pi a^4 G}{8\mu_P}\left[1 - \frac{4}{3}\left(\frac{2\tau_y}{aG}\right) + \frac{1}{3}\left(\frac{2\tau_y}{aG}\right)^4\right] \qquad (2\text{-}83)$$

For $\tau_y = 0$ the expressions reduce to those for Poiseuille flow.

We have given only a sketch here of non-Newtonian flow, and perhaps the most interesting characteristic of these fluids occurs in turbulent flow. Under certain conditions of so-called anomalous turbulent flow the wall friction or head loss in a pipe is greatly reduced so that the effect of turbulence appears suppressed. Only a few non-Newtonian fluids exhibit this characteristic, among them being certain organic polymers and slurries of solids in water such as are used in oil well drilling "muds." We cannot go into a discussion of these interesting properties here, but there is a vast area of engineering application of fluids other than simple Newtonian fluids.

2-13 FIRST INTEGRALS OF THE EQUATION OF MOTION AND BERNOULLI'S EQUATION

Before we continue the applications of the principles of viscous flow to more practical problems, a special useful form of the equations of motion will be derived.

Consider an inviscid fluid. The vector equation of motion is

$$\rho\left[\frac{\partial \mathbf{V}}{\partial t} + \nabla\left(\frac{V^2}{2}\right) - \mathbf{V} \times (\nabla \times \mathbf{V})\right] = -\nabla P - \rho \, \nabla \psi \qquad (2\text{-}84)$$

where ψ is the gravitational potential and $-\rho \, \nabla \psi$ is the conservative gravitational body force per unit volume.* If we take the scalar (dot) product of each term in this equation of motion with a differential elemental length $d\mathbf{l}$ *along a streamline* ($d\mathbf{l}$ is parallel to **V**), then $\mathbf{V} \times (\nabla \times \mathbf{V}) \cdot d\mathbf{l}$ is zero. We integrate the resulting equation (assuming time to be held fixed) between two points (1) and (2) along the streamline. The term $\nabla(V^2/2) \cdot d\mathbf{l}$ is simply $d(V^2/2)$ by the rule of total differentiation, and similarly $\nabla P/\rho \cdot d\mathbf{l}$ is dP/ρ and $\nabla \psi \cdot d\mathbf{l}$ is $d\psi$. Hence

$$\int_1^2 \frac{\partial \mathbf{V}}{\partial t} \cdot d\mathbf{l} + \int_1^2 d\left(\frac{V^2}{2}\right) = -\int_1^2 \frac{dP}{\rho} - \int_1^2 d\psi$$

Now if ρ can be expressed as a function of P (such a fluid is known as *barotropic*), or if ρ is a constant, the integral $\int_1^2 dP/\rho$ can be explicitly evaluated. We have then

$$\int_1^2 \frac{\partial \mathbf{V}}{\partial t} \cdot d\mathbf{l} + \frac{V_2^2 - V_1^2}{2} + \int_1^2 \frac{dP}{\rho} + (\psi_2 - \psi_1) = 0 \qquad (2\text{-}85)$$

as the general form of *Bernoulli's equation.* If the fluid flow is steady and incompressible, the equation is written as

$$\frac{V_2^2 - V_1^2}{2} + \frac{P_2 - P_1}{\rho} + (\psi_2 - \psi_1) = 0 \qquad (2\text{-}86)$$

and $(\psi_2 - \psi_1)$ is often written as $g(z_2 - z_1)$ in engineering problems.

Now, if we wish to write the Bernoulli equation between any two arbitrary points in the fluid, an additional assumption must be made. In general, if $d\mathbf{l}$ and **V** are not parallel, the term $\mathbf{V} \times (\nabla \times \mathbf{V}) \cdot d\mathbf{l}$ is zero only if $\nabla \times \mathbf{V} = 0$, that is, the flow is irrotational. Under this condition Eqs. (2-85) and (2-86) are still valid. A further discussion of the criteria for irrotationality is beyond our needs, but in general the flow will be irrotational if there is no friction (viscosity) in the fluid, or practically speaking it is negligible as in "potential" flow.

Another important and more general form of Bernoulli's equation (sometimes called the *extended Bernoulli equation*) may be obtained if the flow is viscous. The general equation of motion (2-61) may be integrated along a streamline or along a pipe (where the equation is most often used in hydraulics). We shall not show the details, here, but consider a pipe with steady incompressible flow

---

*In introductory fluid mechanics courses, the gravitational body force is often assumed to act in the negative z direction, in which case ψ is gz and $-\rho \, \nabla \psi$ has only a z component equal to $-\rho g$.

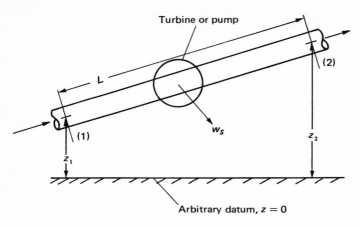

Figure 2-26 A pipe along which the equation of motion is integrated to obtain the extended Bernoulli equation. w_s is positive for a turbine, that is, for power being supplied by the flowing fluid.

with friction and even a pump or turbine (which constitute body forces that are not conservative) in the line (Fig. 2-26). By integrating the equation of motion and taking account of the pressure rise or drop across the pump or turbine, it may be shown that for steady incompressible flow:

$$\frac{V_2^2 - V_1^2}{2} + \frac{P_2 - P_1}{\rho} + g(z_2 - z_1) = -w_s - gH_L \qquad (2\text{-}87)$$

All quantities are based on a unit mass of flowing fluid. Here w_s is the work rate per slug, say, and is positive for a turbine and negative for a pump. H_L is known as the head loss term (due to friction), is often written as $H_L = fLV^2/2Dg$, and has been discussed earlier in Section 2-5. This form in terms of head loss is valid for either viscous or turbulent flow, the friction factor of being a function of the Reynolds number.

For a compressible fluid in steady flow the appropriate expression is simple only in a frictionless fluid (i.e., with no viscosity). Under conditions of no friction the equation of motion integrates to

$$-w_s = \int_1^2 \frac{dP}{\rho} + \frac{V_2^2 - V_1^2}{2} + g(z_2 - z_1) \qquad (2\text{-}88)$$

Later, in Section 4-8, we shall discuss the relationship of this extended Bernoulli equation to the energy equation.

PROBLEMS

2-1 A slow-flow oil drip device designed to feed oil to parts of a machine consists of a tube of diameter D and length L that feeds a short manifold that in turn feeds several small tubes each of diameter d and length l. (There are N of these feed tubes.) See Fig. 2-27. Assume that the flow is slow and laminar and that each small tube empties into a cup reservoir at atmospheric pressure. For a given flow rate q per small tube, what pressure must be developed by the pump?

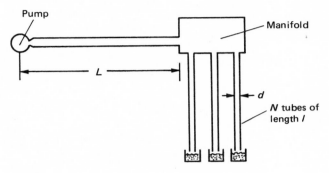

Figure 2-27

2-2 An incompressible fluid flows in an annular pipe as shown in Fig. 2-28. (An annular pipe is formed by inserting a solid cylinder along the axis of a round pipe.) Assuming the flow to be laminar, what is the velocity profile in steady, fully developed flow?

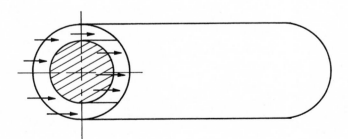

Figure 2-28

2-3 The device illustrated in Fig. 2-29 is known as a viscosity motor. It consists of a stationary case inside of which a drum is rotating. The case and the drum are concentric. Incompressible fluid enters at A, flows through the annulus between the case and the drum, and leaves at B. The pressure at A is higher than at B, the difference being ΔP. The length

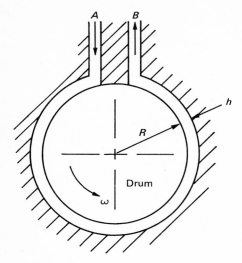

Figure 2-29

of the annulus is l (in the circumferential direction). The width of the annulus h is very small compared to the diameter of the drum, so that the flow in the annulus is equivalent to the flow between two flat plates. Let the density of the fluid be ρ and the viscosity be μ.

(*a*) Find the power delivered to the drum by the fluid and the flow rate as functions of the pressure drop.

(*b*) Find the efficiency in terms of a dimensionless pressure drop.

(*c*) What is the maximum efficiency? (*Answer:* $\frac{1}{3}$.)

(*d*) What happens to the power that is lost?

2-4 A hydraulic dashpot or damper shown in Fig. 2-30 consists of an outer cylinder of inside diameter D into which is placed a solid cylinder of diameter d that is nearly equal to D. The ends of the outer casing cylinder are closed except for a small hole, which allows the shaft to be attached to the solid inner cylinder or plunger. The casing is filled with oil and the shaft is sealed so that oil cannot leak out around it. The outer casing may be

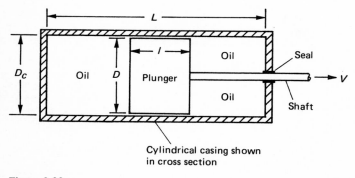

Figure 2-30

fastened rigidly to a mounting and the shaft fastened to a machine member that is to be damped. Automobile shock absorbers work on this principle.

The small clearance between the two cylinders impedes the flow of oil from the front of the plunger to the rear, or vice versa, so that the plunger can only move very slowly. As the plunger moves, the oil must flow around it through the small passage created by the clearance.

Find the resisting force of the dashpot as a function of the relevant parameters and the velocity V at which the plunger is moved.

2-5 Figure 2-31 shows a partial stepped journal bearing that has been proposed as a practical load-supporting device. The bearing is very long and covers a 180° arc, with the step located at the midpoint or 90°. During operation the journal is aligned so that its center is coincident with the centers of the two bearing arcs. If the pressure at the inlet and outlet is atmospheric, find the load W and its direction, and the frictional torque on the journal.

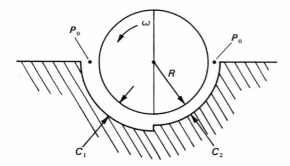

Figure 2-31

2-6 In order to reduce friction in the rolling of wide metal strip, it has been proposed that lubricant guides be positioned to maintain lubricant between the strip and the rolls as it is being reduced (see Fig. 2-32). It is felt that, provided the sealing problems were solved, it might be possible to build up lubricant pressure to the point where it will deform the strip without getting metal-to-metal contact between the strip and the roll, except at the exit point.

(*a*) Assuming that the radial clearance h between the roll and the guide is constant, calculate the length of guide necessary to produce a lubricant pressure of 100,000 psi at the point where the strip is just beginning to be reduced. It may be assumed that the strip and the roll are in contact at the exit plane.

(*b*) Approximately how much power must be delivered by the rolls in this process if the strip is 3 ft wide?

Data:

Radial clearance $h = 0.001$ in.
Fluid viscosity $\mu = 1 \times 10^{-3}$ lb · s/ft²
Incoming strip thickness $= \frac{1}{4}$ in.
Exit strip thickness $= \frac{1}{8}$ in.
Roll diameter $= 12$ in.
Roll speed $= 150$ rpm
Length of strip deformation path $= 0.86$ in. (measured along roll surface)

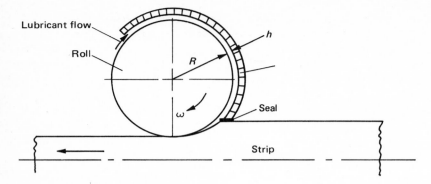

Figure 2-32

2-7 It is believed by some geophysicists that glaciers flow as Newtonian liquids of very high viscosity. As an experiment designed to determine what this viscosity might be, consider a glacier flowing slowly but steadily down an inclined plane surface of rock that makes a known angle with the horizontal. Measurements can be made of the density of the glacier, its thickness, and the velocity of its surface. Develop an expression for the viscosity of the glacier in terms of these parameters.

2-8 Fully developed steady flow between two flat, very wide, porous plates takes place in the presence of a cross flow. Fluid flows with a uniform cross velocity V_0 between the two plates, which are porous. (See Fig. 2-33). Determine the velocity profile in the channel.

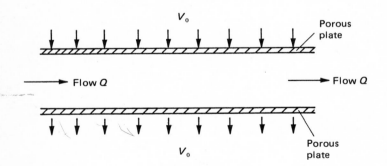

Figure 2-33

2-9 Some painters believe that paint is a Bingham plastic because when they paint a vertical wall, the paint doesn't all run off. The amount of paint that remains on a vertical surface depends on the parameter τ_y, μ_p, and the rate at which the paint dries. Assuming that drying occurs slowly after the paint thickness has been established, find an expression for the thickness of the paint layer on a vertical surface.

2-10 A Bingham plastic is confined between two plates. What force is necessary to set the plates in relative motion? Sketch the velocity profile after motion begins.

2-11 A Bingham plastic flows between large parallel stationary plates. Find the velocity profile and flow rate in terms of the pressure gradient and other relevant parameters.

2-12 Solve Problem 2-11 for a power-law fluid.

2-13 A viscous liquid is contained between long concentric cylinders. Each cylinder rotates about its axis with a steady arbitrary angular velocity. Determine the velocity profile in the liquid. Discuss the profile for a Bingham plastic.

2-14 A rotating viscosimeter is shown in Fig. 2-34. It consists of a rotating screw with tight-fitting rectangular threads set in a cylinder. One end of the screw is open to the liquid and the other end is closed off with a pressure gauge attached. A relation among the pressure reading, viscosity, screw geometry, and screw speed is desired.

Such a screw can also be used as a shaft seal. Discuss such usage and comment on the relevant parameters for seal operation.

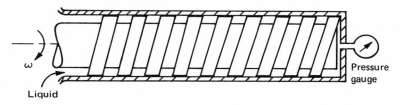

Figure 2-34

2-15 Steady laminar air flow occurs in a round tube of diameter D and length L. The pressure at the inlet and outlet are P_1 and P_2, respectively. Assuming that the flow is isothermal (a good assumption if the walls are good thermal conductors), find the pressure variation along the tube.

2-16 Show that if steady laminar flow occurs in a round tube that is inclined to the horizontal, the results are identical to those presented for a horizontal tube in the text with the pressure P replaced by $(P - \rho gh)$, where h is the elevation above some specified datum level. Explain how this datum is defined in terms of some known pressure such as that produced by a pump.

BIBLIOGRAPHY

General

Bird, R. B., W. E. Stewart, and E. N. Lightfoot: "Transport Phenomena," chap. 2, John Wiley & Sons, New York, 1960.

Hughes, W. F., and E. W. Gaylord: "Basic Equations of Engineering Science," Schaum Outline Series, McGraw-Hill Book Company, New York, 1964.

Shames, I. H.: "Mechanics of Fluids," McGraw-Hill Book Company, New York, 1962.

Non-Newtonian Flow

Middleman, S: "The Flow of High Polymers: Continuum and Molecular Rheology," Wiley Interscience Publishers, New York, 1968.

Skelland, A. H. P.: "Non-Newtonian Flow and Heat Transfer," John Wiley & Sons, New York, 1967.

Streeter, V. L. (ed.): "Handbook of Fluid Dynamics," chap. 7, by A. B. Metzner, McGraw-Hill Book Company, New York, 1961.

Wilkinson, W. L.: "Non-Newtonian Fluids," Pergamon Press, Elmsford, N.Y., 1960.

HYDRODYNAMIC LUBRICATION

3-1 INTRODUCTION

In everyday experience we tend to think of a surface being lubricated if it is "slippery." If we try to define a slippery surface, we might say that it is one that exhibits little friction or resistance to an object sliding over it. These intuitive concepts are indeed correct, but the origin of the slippery character of a surface may not be so simple to understand. A surface may be slippery or, in more precise terms, have a low coefficient of friction (with respect to a given object sliding over it) because of the chemical composition of the surface. An example is the synthetic plastic Teflon, which has very low static and dynamic coefficients with respect to itself and most other materials. A hard surface may also be coated with a powder such as graphite, which tends to act in a similar fashion and reduce friction.

Many such dry bearings, which rely on the material itself to reduce the coefficient of friction, have been developed in the past few years and used extensively in all sorts of devices. However, many materials such as Teflon wear rapidly, tend to deform plastically, and are difficult to keep in precise alignment.

Since the wheel was invented, and maybe before, the principle of hydrodynamic lubrication has been used, although it is only in this century that it has been really understood. A film of lubricant, oil, water, air, grease, or other fluid may be used to reduce the wear that would result if the surfaces were dry. The fluid keeps the surfaces apart, and the friction that results is due only to viscous shear in the fluid, which is usually many times less than the dry friction. The lubricating property of the fluid is not due to its chemical or "slippery" nature but it is the consequence of simple viscous (laminar) flow of the fluid between the surfaces. Lubrication by means of a fluid is known as *hydrodynamic lubrication* and is of vital importance in just about everything that moves. Whenever relative motion takes place between two objects, lubrication is necessary. Every joint in the human body is lubricated by a natural fluid, and when the fluid becomes ineffective arthritis results. Every mechanical device or machine that has at least one moving part must be lubricated. It is rather obvious, then, that some knowledge of hydrodynamic lubrication is essential to an engineer. And here we find one of the most important applications of viscous flow theory.

There are many types of bearings that have been devised for various purposes—from huge bearings on power shafts that support many tons to small, high-speed, precision gas bearings used in gyroscopes. The basic principle of all these bearings is the same: The fluid has a pressure buildup in it that supports the load. There are two general classes of bearings, (1) dynamic or self-acting and (2) hydrostatic.

The dynamic bearing becomes pressurized and able to support a load because of its motion and consequent forcing of the lubricating fluid between the surfaces by the motion of the surfaces themselves. Such bearings are useful for both small and heavy loads where motion is always taking place at a sufficiently high speed.

The hydrostatic or pressurized bearing relies on an external pump or pressurization system to maintain a high pressure between the mating surfaces. The pressure and load-supporting capacity can be maintained even under zero relative speed of the mating surfaces. Pressurized bearings are useful for slow speed and in situations where precise separation under varying conditions must be maintained. Often the two systems are used together, the dynamic effect supplementing a background or feed pressure.

In Fig. 3-1 we show a few simple types of bearings. The inclined pad slider moves over the bearing and the fluid moves in a Couette-type flow between the surfaces. Since the surfaces are tilted the fluid tends to build up a pressure as it flows through the channel, and this pressure can support the slider pad. Most dynamic bearings are a variation of the inclined pad. The step slider and the journal bearing, which looks like an inclined slider wrapped around a cylinder, are two commonly used examples of a dynamic or self-acting bearing. A simple hydrostatic thrust bearing is also shown. An external pump keeps fluid at high pressure flowing between the surfaces regardless of the relative motion of the surfaces.

In this chapter we shall analyze these simple bearings in detail, calculating the pressure distribution in the lubricating film, the friction on the surfaces, and

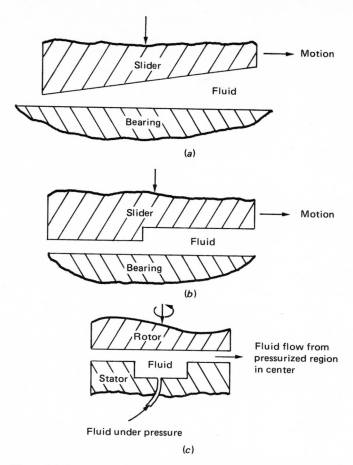

Figure 3-1 Some general types of bearings in use. (*a*) Inclined slider. (*b*) Step slider. (*c*) Hydrostatic thrust. The pressure builds up if the fluid is viscous as it flows through the tapered channel. It is this pressure that supports the load.

the load-carrying capacity. We shall idealize the flow to a certain extent, but the results will correspond closely to actual performance for a large class of practical bearings. Then we shall discuss a few of the more sophisticated corrections that may be made to the basic theory. In engineering practice, for the purposes of design, extensive tables have been constructed that show the results of theoretical calculations (some obtained with digital computers) and experimental data. In the design of new types of bearings for special purposes, however, it is often necessary to return to basic theory and examine the behavior of the lubricating fluid in detail.

If the wheel was the first and perhaps most significant invention in the history of mankind, it certainly follows that the use of lubricants in man-made machines must have occurred very early; indeed, without the knowledge that a heavy-bodied (or viscous) fluid could reduce friction, it is unlikely that any inventions beyond the wheel would have been possible.

Before we begin the detailed study of some particular bearings, there are some general characteristics of lubrication flow that one can mention.

1. The spacing between the slider and bearing surfaces is usually small (of the order of 10^{-3} in.) compared to the length of the slider (of the order of inches). Hence the Reynolds number based on the spacing is usually small (much less than unity), and the inertia effects in the flow are small and can be neglected to a good approximation.
2. The development length in the flow channel is much smaller than the length of the channel, and the flow is usually considered to be fully developed.
3. The flow is usually laminar.

Under extreme conditions any or all of these assumptions break down and a more complex analysis is required. Later in this chapter we shall briefly mention the effect of including inertia in the calculations.

3-2 THE STEP SLIDER BEARING

We shall begin the study of bearings by considering a bearing of simple geometry, the step slider bearing. The basic mathematical formulation and underlying physical assumptions will be discussed in relationship to this particular bearing, and later generalized to other types of bearings.

Referring to Fig. 3-2, we consider the two-dimensional step slider moving with velocity U with respect to the flat bearing surface. Generally it is convenient to pick a coordinate system fixed to the slider. Then the surface of the slider that usually contains the irregularities is at rest with respect to the coordinate system. Further, it is convenient to place the origin of the coordinate system on the surface of the bearing and allow one coordinate axis (x here) to be parallel to the bearing surface. In this way the boundary conditions and velocity will be simplified because the slider that contains the irregularities in shape is at rest and the bearing surface moves in a single coordinate direction. For the step slider it is convenient to use two coordinate systems as shown in Fig. 3-2.

If we neglect the development region in the flow channel, the step bearing

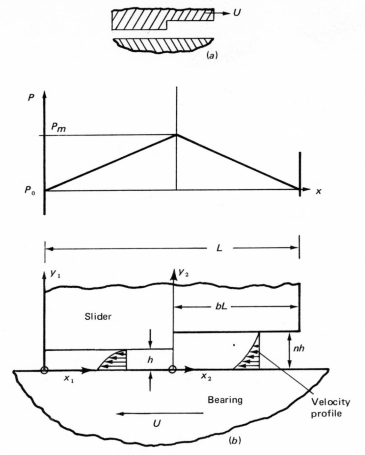

Figure 3-2 The two-dimensional step slider bearing. The actual motion is shown in (*a*). The coordinate system is affixed to the slider as shown in (*b*) and then the bearing moves with velocity U in the negative x direction with respect to the slider.

reduces exactly to the problem of two matched Couette flow problems. The flow rate in the two chambers of depth nh and h must be equated by continuity to find the pressure at the step. The two-dimensional character of the flow in the region of the step is ignored because it will be a localized effect since the spacings are so small.

In the region of spacing h, the equation of motion is

$$0 = -\frac{\partial P}{\partial x_1} + \mu \frac{\partial^2 u}{\partial y^2} \tag{3-1}$$

with the boundary conditions $u = 0$, $y = h$; and $u = -U$, $y = 0$. Following the method of Chapter 2, we find

$$u = \frac{1}{2\mu}\frac{dP}{dx_1}(y^2 - hy) + U\left(\frac{y}{h} - 1\right)$$

(3-2)

and the flow rate Q as

$$Q = -\frac{h^3}{12\mu}\frac{dP}{dx_1} - \frac{Uh}{2} = -\frac{h^3}{12\mu}\left[\frac{P_m - P_0}{L(1-b)}\right] - \frac{Uh}{2}$$

(3-3)

where P_0 is the entrance and exit pressure, usually atmospheric, and P_m is the pressure on the step. Similarly, in the region where the spacing is nh, the velocity profile and flow rate are

$$u = \frac{1}{2\mu}\frac{dP}{dx_2}(y^2 - nhy) - U\left(\frac{y}{nh} - 1\right)$$

(3-4)

and

$$Q = -\frac{(nh)^3}{12\mu}\left(\frac{P_0 - P_m}{bL}\right) - \frac{Unh}{2}$$

(3-5)

Equating the two expressions for Q, we find P_m as

$$P_m - P_0 = \frac{6U\mu Lb(n-1)(1-b)}{h^2[b + n^3(1-b)]}$$

(3-6)

Then the total vertical load W (per unit width of channel) that the slider can support is simply the integral of the pressure or the area under the pressure versus x curve shown in Fig. 3-2.

$$W = (P_m - P_0)\frac{L}{2} = \frac{3U\mu L^2 b(n-1)(1-b)}{h^2[b + n^3(1-b)]}$$

(3-7)

It is convenient to define a dimensionless load coefficient C_w such that $W = C_w(U\mu L^2/h^2)$ and then

$$C_w = \frac{Wh^2}{U\mu L^2} = \frac{3b(n-1)(1-b)}{b + n^3(1-b)}$$

(3-8)

The question now arises as to what is the optimum value of n and b to get the maximum value of W for a given fluid and speed U. We shall not carry out the details here, but the results follow from a straightforward calculation and are shown in Table 3-1 for both the infinite slider considered here and for a square step slider of dimensions $L \times L$. The effect of making the slider finite in width (out of the paper in Fig. 3-2) is to allow leakage out the sides with a consequent reduction in load-carrying capacity. The calculations for a finite-width slider

Table 3-1 Values of C_W for Step and Slider Bearings of Optimum Step Size and Location

	Infinite Step Slider	Square Step Slider	Infinite Inclined Slider	Square Inclined Slider
b	0.721	0.545	—	—
n	1.87	1.70	2.19	2.0
C_W	0.206	0.072	0.160	0.072

require a Fourier series solution to a partial differential equation and are too specialized to pursue here.

The frictional drag can now be computed for the infinitely wide step slider. We can integrate the shear stress either over the surface of the bearing or over the surface of the slider. In the latter case we must add the pressure force on the step $(P_m - P_0)h(1 - n)$ per unit width of slider. Either method of finding the drag force will give the same result.

Evaluating the shear τ_w on the bearing surface $(y = 0)$, we find

$$\tau_{w_1} = \mu\left(-\frac{h}{2\mu}\frac{dP}{dx_1} + \frac{U}{h}\right)$$

in the region of spacing h, and in the region of spacing nh we find

$$\tau_{w_2} = \mu\left(-\frac{nh}{2\mu}\frac{dP}{dx_2} + \frac{U}{nh}\right)$$

Using the value of P_m from Eq. (3-6), we find for the drag D on the slider,

$$D = \int_0^{L(1-b)} \tau_{w_1}\,dx_1 + \int_0^{bL} \tau_{w_2}\,dx_2 = \frac{3LU\mu b(n-1)^2(1-b)}{h[b + n^3(1-b)]}$$

$$+ \frac{LU\mu}{h}\left[\frac{b + n(1-b)}{n}\right] \quad (3\text{-}9)$$

A dimensionless friction coefficient C_f may be defined as D/W, and for the step slider we find

$$C_f = \frac{h(n-1)}{L} + \frac{h[n(1-b) + b][b + n^3(1-b)]}{3Lbn(n-1)(1-b)} \quad (3\text{-}10)$$

3-3 THE GENERAL MATHEMATICAL FORMULATION

Before we extend our consideration to a slider of arbitrary shape, it is convenient to consider the mathematical model for a general dynamical slider bearing.

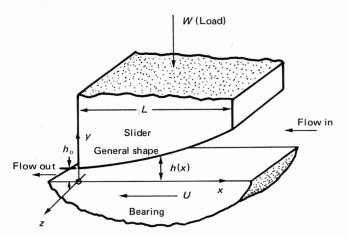

Figure 3-3 The general hydrodynamic slider bearing. The drawing is not to scale since $L \gg h$. The coordinate system is attached to the slider.

First we consider the two-dimensional slider bearing shown in Fig. 3-3. The spacing h (due to the curvature of the slider) is considered to be an arbitrary function of x. Again we consider a coordinate system attached to the slider but with the origin coincident with the bearing surface.

Let us write the equations of motion of the lubricant and make an order-of-magnitude estimation of the various terms. For incompressible, steady flow, the equations of motion and continuity are

$$x: \quad \rho\left(u \frac{\partial u}{\partial x} + v \frac{\partial u}{\partial y}\right) = -\frac{\partial P}{\partial x} + \mu\left(\frac{\partial^2 u}{\partial x^2} + \frac{\partial^2 u}{\partial y^2}\right)$$

$$y: \quad \rho\left(v \frac{\partial v}{\partial x} + v \frac{\partial v}{\partial y}\right) = -\frac{\partial P}{\partial y} + \mu\left(\frac{\partial^2 v}{\partial x^2} + \frac{\partial^2 v}{\partial y^2}\right)$$

$$\frac{\partial u}{\partial x} + \frac{\partial v}{\partial y} = 0$$

We begin by nondimensionalizing the x equation of motion in order to facilitate estimation of the relative size of the terms. We let nondimensional variables be

$$x' = \frac{x}{L}$$

$$y' = \frac{y}{h_0}$$

$$u' = \frac{u}{U}$$

$$v' = \frac{v}{U}\frac{L}{h_0}$$

$$P' = \frac{P}{\rho U^2}$$

$$\mathrm{Re} = \frac{\rho U h_0}{\mu}$$

which leads directly to

$$\left(u'\frac{\partial u'}{\partial x'} + v'\frac{\partial u'}{\partial y'}\right) = -\frac{\partial P'}{\partial x'} + \frac{1}{\mathrm{Re}}\left(\frac{L}{h_0}\right)\left(\frac{h_0^2}{L^2}\frac{\partial^2 u'}{\partial x'^2} + \frac{\partial^2 u'}{\partial y'^2}\right) \qquad (3\text{-}11)$$

In lubrication flow the Reynolds number Re based on h_0 will usually be close to unity and hence the effective Reynolds number $\mathrm{Re}(h_0/L)$ will be smaller by a factor of approximately 10^{-3}. We conclude that the inertia effects are generally small and the pressure and viscous forces must balance. For high-speed flow the inertia terms may become important, and approximate methods for taking them into account will be discussed later. However, the viscous term $\partial^2 u'/\partial x'^2$ is multiplied by $(h_0/L)^2$ and will always be negligible compared to $\partial^2 u'/\partial y'^2$.

Taking these assumptions into account, Eq. (3-11) becomes simply

$$0 = -\frac{\partial P}{\partial x} + \mu\frac{\partial^2 u}{\partial y^2} \qquad (3\text{-}12)$$

which is identical to the equation we have used extensively for flow between parallel plates. Physically, the slight degree of tilt of a bearing allows a quasi-one-dimensional analysis and the fact that h is a function of x enters only in the continuity equation.

Referring to this Eq. (3-12), it may be solved along with the continuity equation to give $u(x, y)$ and $P(x)$ if the viscosity μ is assumed constant. Because of the viscous dissipation in the lubricant film, the temperature may rise in the fluid and change the viscosity as the fluid flows through the channel. If this change becomes important, the energy equation must be solved simultaneously with motion and continuity. And if the lubricant is compressible (a gas), the energy equation in one form or another (or an assumption that the flow is isothermal) must be used even if μ is assumed constant. For most gas bearings the assumption of isothermal flow is not too far from actuality. For liquids, however, the temperature may rise considerably through the channel, and more sophisticated models take this into account in terms of viscosity changes.

The key to solving Eq. (3-12) is to realize that h is a slowly varying function of x and Eq. (3-12) may be integrated with respect to y assuming h to be constant. Hence

$$u = \frac{1}{2\mu}\frac{dP}{dx}(y^2 - hy) + U\left(\frac{y}{h} - 1\right) \tag{3-13}$$

and, by continuity,

$$Q = \int_0^{h(x)} u\,dy = -\frac{h^3}{12\mu}\frac{dP}{dx} - \frac{Uh}{2} = \text{constant} \tag{3-14}$$

But now h must be treated as a function of x and hence Eq. (3-14) is a differential equation for $P(x)$, given $h(x)$. The two boundary conditions (that $P = P_0$ at $x = 0, L$) serve to determine Q and the one constant of integration.

If Eq. (3-14) is differentiated with respect to x, we find that $dQ/dx = 0$ and

$$\frac{d}{dx}\left(\frac{h^3}{\mu}\frac{dP}{dx}\right) + 6U\frac{dh}{dx} = 0 \tag{3-15}$$

which is a fundamental equation of lubrication theory known as the *Reynolds equation*, which is valid even for variable viscosity. (For a coordinate system in which the direction of positive x is reversed from the one we are using, the sign preceding the second term becomes a negative sign. This form is often seen in the literature.)

The Reynolds equation can be generalized to two dimensions for a finite pad slider (finite in the z direction) in which h is a function of x and z:

$$\frac{\partial}{\partial x}\left(\frac{h^3}{\mu}\frac{\partial P}{\partial x}\right) + \frac{\partial}{\partial z}\left(\frac{h^3}{\mu}\frac{\partial P}{\partial z}\right) + 6U\frac{\partial h}{\partial x} = 0 \tag{3-16}$$

which is a partial differential equation and will not be discussed further here, although in practice the side leakage predicted by the finite size of the slider is of utmost importance.

3-4 THE INCLINED SLIDER BEARING

The two-dimensional inclined slider bearing shown in Fig. 3-4 is the simplest type of dynamical bearing, other than the step slider, and nearly all practical slider bearings are a variation of this basic slider bearing. The film thickness variation $h(x)$ may be written as

$$h = h_0\left[1 + (n-1)\frac{x}{L}\right] \tag{3-17}$$

The velocity profile may be immediately obtained by substituting the above

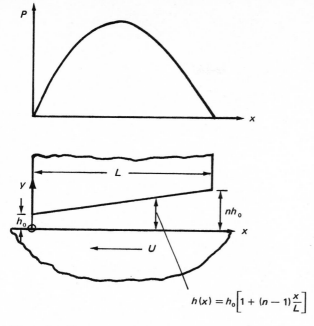

$$h(x) = h_0\left[1 + (n-1)\frac{x}{L}\right]$$

Figure 3-4 The two-dimensional inclined slider bearing. (Not to scale, since $L \gg h$.)

expression for h into Eq. (3-13) and the equation for pressure $P(x)$ by substituting into Eq. (3-15) or Eq. (3-14) and using the boundary conditions that $P = P_0$ at $x = 0, L$. Integrating, we find for $P(x)$,

$$P(x) - P_0 = \frac{\mu UL}{h_0^2} \frac{6(n-1)[1-(x/L)](x/L)}{(n+1)[n+(n-1)x/L]^2} \tag{3-18}$$

which is very nearly a parabolic distribution with the maximum pressure occurring nearer the trailing edge.

The load coefficient C_w [where $W = (\mu UL^2/h_0^2)C_w$] is

$$C_w = \frac{6}{(n-1)^2}\left[\ln n - \frac{2(n-1)}{n+1}\right] \tag{3-19}$$

The optimum value of n and the corresponding value of C_w are shown in Table 3-1. The coefficient of friction C_f is found to be

$$C_f = \frac{h_0(n-1)}{L}\left[\frac{\ln n}{6 \ln n - 12(n-1)/n + 1} + \frac{1}{2}\right] \tag{3-20}$$

3-5 THE HYDROSTATIC THRUST BEARING

The simple hydrostatic thrust bearing shown in Fig. 3-5 can support a load even under no-motion conditions. The lubricant is supplied under pressure and fills the recess region at essentially the supply pressure P_i. The lubricant then slowly flows out between the rotor and stator as it drops to atmospheric pressure P_0.

Such bearings are useful for supporting very large thrust loads of rotating machinery even under low-speed conditions. The rotation of the stator has a very

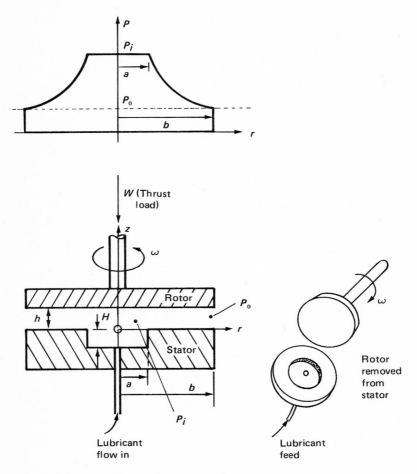

Figure 3-5 The simple hydrostatic thrust bearing. The recess region is of depth H and radius a. H is usually made much larger than h so that the pressure is nearly uniform throughout the recess.

small effect on the load capacity, which is determined mainly by the supply pressure that can be maintained in the recess.

For a fixed flow rate of lubricant the pressure P_i depends on the film thickness (h). Hence if a given pump with a fixed flow rate is used, the film thickness will decrease as the load W is increased and vice versa. If, on the other hand, a given film thickness must be maintained regardless of the load W, then the flow rate must be increased as the load is increased.

The design problem, then, is to relate the flow rate of lubricant Q to the film thickness (h) and the load W.

In the analysis here we shall assume the lubricant incompressible, the viscosity constant, and the effects of inertia will be neglected. At high rotational speed the centrifugal inertia terms may become important and affect the load capacity. We shall discuss this effect in the next section.

The appropriate differential equation for the radial flow between the rotor and stator may be obtained as we did for the dynamic pad bearings, but we must work in cylindrical coordinates. We can simply look up the appropriate forms of the Navier-Stokes equation, or we can derive the equations from fundamentals.

Since we have not derived this equation yet, we shall derive the radial component by considering an elemental volume. Consider the pie-shaped element in Fig. 3-6. Neglecting the inertia terms, we sum the pressure and viscous shear forces in a vector direction. Note that we do not sum forces in simply the r direction. We pick a vector direction that coincides with the r direction only along the center line of the element. For radially symmetrical flow there are shear forces only on the top and bottom (which act in the vector r direction). Note that the pressure forces on the sides of the element, while acting in the tangential (or θ direction) have components in the vector r direction. This is an important point often confusing at first.

Summing the forces, we have (remembering that $\sin \Delta\theta/2 \approx \Delta\theta/2$ since $\Delta\theta$ is small; u is the component of velocity in the r direction)

$$-P|_{r+\Delta r}(r + \Delta r)\,\Delta\theta\,\Delta z + P|_r r\,\Delta\theta\,\Delta z + P|_{r + \epsilon\Delta r}\Delta r\,\Delta z\,\Delta\theta$$

$$+ \mu\frac{\partial u}{\partial z}\bigg|_{z+\Delta z}(r + \epsilon\,\Delta r)\,\Delta r\,\Delta\theta - \mu\frac{\partial u}{\partial z}\bigg|_z(r + \epsilon\,\Delta r)\,\Delta r\,\Delta\theta = 0 \quad (3\text{-}21)$$

Taking the appropriate limit, we obtain

$$0 = -\frac{\partial P}{\partial r} + \mu\frac{\partial^2 u}{\partial z^2} \qquad (3\text{-}22)$$

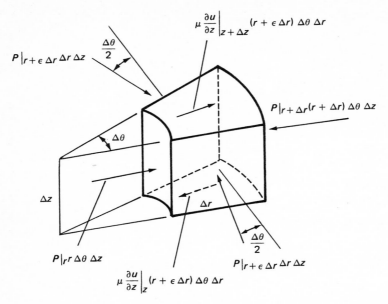

Figure 3-6 An elemental volume for derivation of the radial equation motion in cylindrical coordinates.

which is just like the equation in Cartesian coordinates. (However, the inertia terms would be quite different and the other shear terms would have been different.) Note here that there are additional shear terms that come about when we allow a general linear relationship between stress and strain rates, as was discussed in Section 2-10. These additional viscous terms are small in this problem, and their exact form may be studied by referring to the complete equation of motion in cylindrical coordinates.

The equation in the tangential direction is simply

$$\frac{\partial^2 v}{\partial z^2} = 0 \tag{3-23}$$

where v is the θ component of velocity. This equation may be integrated at once with the boundary conditions $v = 0, z = 0; v = r\omega, z = h$ to give

$$v = r\omega \frac{z}{h}$$

which, if we neglect inertia effects, is needed only for determination of the frictional torque and not the pressure distribution.

Now, Eq. (3-22) may be integrated with the boundary conditions $u = 0$, $z = 0, h$. We assume that the pressure is essentially constant in the recess region since $H \gg h$ and the velocity there is negligible. We find in the flow region $(b > r > a)$:

$$u = \frac{1}{2\mu} \frac{dP}{dr} (z^2 - hz) \qquad (3\text{-}24)$$

Continuity gives the radial flow rate Q as

$$Q = \int_0^h 2\pi r u \, dz = -\frac{\pi r h^3}{6\mu} \frac{dP}{dr}$$

As before, P is a function only of r, and using the boundary conditions $P = P_i$ at $r = a$ and $P = P_0$ at $r = b$, we obtain for $P(r)$ and Q,

$$Q = \frac{\pi h^3 (P_i - P_0)}{6\mu \ln (b/a)} \qquad (3\text{-}25)$$

$$\frac{P_i - P}{P_i - P_0} = \frac{\ln (r/a)}{\ln (b/a)} \qquad (3\text{-}26)$$

The pressure distribution is explicitly independent of Q or h and depends purely on geometry and the inlet and outlet pressures. Implicitly, however, the pressure P_i is interrelated to the load, flow rate, and film thickness.

The load W is found next as

$$W = \int_a^b (P - P_0)2\pi r \, dr + (P_i - P_0)\pi a^2$$

$$= \frac{\pi (b^2 - a^2)(P_i - P_0)}{2 \ln (b/a)} \qquad (3\text{-}27)$$

again depending only on geometry and the inlet pressure P_i. We might ask why we do not make a very close to b (the recess covering most of the bearing and the flow region is made only as a "lip") so that the load can be made larger for a given P_i. The answer is that the flow rate Q would then become large (the lip offering little viscous resistance to flow). From a practical standpoint the flow region size and film thickness must be designed to give the required load with an acceptable flow rate and associated feed pump.

3-6 THE EFFECT OF INERTIA IN LUBRICATION

So far we have neglected the inertia terms in the equation of motion for lubrication flow. These terms are generally nonlinear and quite difficult to handle. When they are included, they show that the load-carrying capacity is actually somewhat less than would be predicted by the simpler theory in which inertia was neglected.

The inertia effects may at least be considered approximately in the hydrostatic thrust bearing discussed in the previous section. The complete equations of motion in the r and θ directions may be derived by including the momentum fluxes through the element of Fig. 3-6 (just as we did when we derived the equation of motion in Cartesian coordinates).

The equation of motion in the r direction is for axially symmetric flow

$$\rho\left(u\frac{\partial u}{\partial r} - \frac{v^2}{r}\right) = -\frac{\partial P}{\partial r} + \mu\frac{\partial^2 u}{\partial z^2} \tag{3-28}$$

and in the θ direction,

$$\rho\left(u\frac{\partial v}{\partial r} + \frac{uv}{r}\right) = \mu\frac{\partial^2 v}{\partial z^2} \tag{3-29}$$

where again we have neglected the viscous terms involving derivatives in the r direction. P is essentially a function only of r.

For high rotational speeds v will be much larger than u, and we can neglect $u(\partial u/\partial r)$ compared to v^2/r in Eq. (3-28) and neglect both inertia terms (the left-hand side) of Eq. (3-29). In other words, we are including the centrifugal effects due to rotation but are neglecting the acceleration terms due to changes of u with respect to r.

As before, Eq. (3.29) gives $v = r\omega(z/h)$, and we can insert this value for v directly into Eq. (3-28), giving

$$-\frac{\rho r\omega^2 z^2}{h^2} = -\frac{dP}{dr} + \mu\frac{\partial^2 u}{\partial z^2} \tag{3-30}$$

which can be integrated (with the condition that $u = 0$ at $z = 0, h$) to give the velocity profile

$$u = \frac{1}{\mu}\frac{dP}{dr}\frac{z(z-h)}{2} - \frac{\rho r\omega^2}{\mu h^2}\frac{z(z^3 - h^3)}{12} \tag{3-31}$$

From continuity,

$$m = \rho Q = 2\pi r\rho\int_0^h u\,dz$$

from which we obtain

$$\frac{dP}{dr} = -\frac{6m\mu}{\pi\rho r h^3} + \frac{3\rho r \omega^2}{10}$$

Using $P = P_i$ at $r = a$ and P_0 at $r = b$, we integrate for $P(r)$:

$$\frac{P_i - P}{P_i - P_0} = \frac{\ln (r/a)}{\ln (b/a)} - \frac{3\rho(a\omega)^2}{20(P_i - P_0)} \left\{ \left(\frac{r}{a}\right)^2 - 1 - \left[\left(\frac{b}{a}\right)^2 - 1\right] \frac{\ln (r/a)}{\ln (b/a)} \right\} \qquad (3\text{-}32)$$

The load capacity may be calculated as in the previous section and expressed as a ratio of load with inertia to load without inertia:

$$\frac{W}{W_0} = 1 - k \frac{\rho(a\omega)^2}{P_i - P_0} \qquad (3\text{-}33)$$

where W_0 is the load given by Eq. (3-27) and k is a geometrical parameter,

$$k = \frac{3}{20} \left\{ \left[\left(\frac{b}{a}\right)^2 + 1\right] \ln \left(\frac{b}{a}\right) - \left[\left(\frac{b}{a}\right)^2 - 1\right] \right\} \qquad (3\text{-}34)$$

Note that the inertia effect is proportional to the square of the speed and is independent of viscosity.

As a simple example, consider a bearing with $a = 1$ in., $b = 3$ in., $P_i = 3$ atm, and an oil with a density (ρg) of 0.032 lb/in.^{3} (0.887 g/cm^3). For a speed of 2700 rpm, we find that the error in load capacity due to neglecting inertia [from Eq. (3-33)] is about 10%.

We conclude that inertia effects are quite important in rotating bearings of this type, at least for liquid lubricants. Gases (particularly air) are used quite extensively for hydrostatic bearings (of rotating and pad type) because of their low friction, cleanliness, and because the gas need not be recycled as oil must. The inertia effects in gas lubricants are much smaller than for liquid lubricants (for a given load capacity and speed) simply because of the lower density of gases.

3-7 THE JOURNAL BEARING

The journal bearing shown in Fig. 3-7 is used extensively in rotating machinery. The journal or shaft rotates in the fixed bearing and because the journal is made with a slightly smaller diameter than the bearing the rotating motion causes the journal to "ride" to one side of the bearing. This eccentricity gives rise to a film thickness $h(\theta)$, which is in effect an approximation to that of an inclined slider

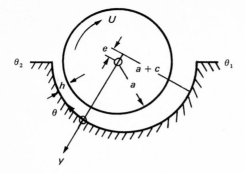

Figure 3-7 A partial journal bearing. The inlet and outlet are located at θ_1 and θ_2, respectively, where the pressure is zero gauge. We can use x or θ as the position coordinates, where $x = a\theta$. The radius of the bearing is $a + c$, where c is the clearance.

bearing. Consequently, a supporting pressure is built up in the film under conditions of rotation, and a load may be supported by the journal.

Journal bearings may be made with a partial bearing, as shown in the 180° bearing of Fig. 3-7, or they may be made with a full bearing that extends 360° around the journal. The analysis of a partial or full journal bearing is the same except for the boundary conditions on pressure. In a partial bearing the pressure is zero (gauge) at the inlet and outlet, but in a full journal an oil inlet hole (which should be located at some optimum position around the bearing) determines the boundary conditions on pressure. Most journal bearings are self-acting, that is, dynamical bearings, but lubricant may be supplied under pressure through an inlet hole on the bottom to augment the load capacity with a hydrostatic pressure.

Because the film thickness $h(\theta)$ is small compared to the journal radius a, the bearing may be unwrapped and the problem treated as a slider bearing in Cartesian coordinates, the only difference being, as we shall see, in the boundary conditions on pressure. The film thickness in a journal bearing decreases from the inlet to a minimum value and then, unlike the usual slider bearing, increases again toward the outlet. The divergence of the film can give rise to cavitation in the film if the lubricant is a liquid. The pressure in a liquid cannot normally drop below zero absolute (in fact, not below the vapor pressure), and when it reaches this value the fluid will separate into fingers or froth. This behavior is known as *cavitation*.

Referring to Fig. 3-7, the journal has radius a, and the bearing has radius $a + c$, where c is the clearance. When the journal and bearing are concentric, the film thickness is simply c. Under load conditions the center of the journal is displaced a distance e. In terms of the eccentricity ratio ϵ, defined as e/c, the film thickness h may be expressed very accurately, but not quite exactly, as

$$h = c(1 - \epsilon \cos \theta) \tag{3-35}$$

where θ is measured from the point of minimum film thickness. The bearing and journal may be unwrapped and the journal assumed to move with respect to the bearing. The coordinate system is attached to the stationary bearing. We assume no side leakage, that is, infinite width of the journal. The equation of motion (3-12) may be integrated with the boundary conditions $y = 0$, $u = U = a\omega$; $y = h$, $u = 0$ to give the velocity profile and then continuity used to give the equation for $dP/d\theta$ in terms of the flow rate Q as

$$\frac{dP}{d\theta} = \frac{6\mu Ua}{h^2} - \frac{12\mu aQ}{h^3} \tag{3-36}$$

This equation may be integrated to find the pressure variation with θ by using the appropriate boundary conditions to find the constant of integration and the flow rate Q. At the inlet $\theta = \theta_1$, the pressure is zero (gauge) but the condition that the pressure is zero at the outlet $\theta = \theta_2$ is, in general, inadequate. If this latter condition is used, the solution may predict a pressure drop to below zero (absolute or the vapor pressure of the liquid) at some location along the film. Actually, this boundary condition of zero gauge pressure at $\theta = \theta_2$ has been used and is known as the *Sommerfeld condition*. This is the condition used by Arnold Sommerfeld in 1904 when he first analyzed the journal bearing. Obviously the solution breaks down where the pressure drops below zero, and it is assumed that the fluid cavitates at that position along the film and remains zero from there to the outlet.

However, once the fluid cavitates, the equations of motion no longer hold and the prediction of the point of cavitation by using an exit boundary condition is not meaningful. To overcome this difficulty the "Reynolds" boundary condition is generally used in present-day calculations. If the pressure remains above zero, using the Sommerfeld condition, the solution is correct. If the pressure drops below zero, then the Reynolds condition is that the pressure must go to zero where the pressure gradient goes to zero. That is,

$$P = 0$$

where $dP/d\theta = 0$, and the value of Q must be determined by trial and error to satisfy this condition.

Equation (3-36) may be integrated to satisfy the first condition that $P = 0$ at $\theta = \theta_1$ as

$$P(\theta) = \frac{6\mu Ua}{c^2} J_2 - \frac{12\mu aQ}{c^3} J_3 \tag{3-37}$$

where J_2 and J_3 are the "Sommerfeld" integrals, defined as

$$J_2 = \int_{\theta_1}^{\theta} \frac{d\theta}{(1 - \epsilon \cos \theta)^2} = \frac{(\gamma - \gamma_1) + \epsilon(\sin \gamma - \sin \gamma_1)}{(1 - \epsilon^2)^{3/2}}$$

$$J_3 = \int_{\theta_1}^{\theta} \frac{d\theta}{(1 - \epsilon \cos \theta)^3} \tag{3-38}$$

$$= \frac{(1 + \epsilon^2/2)(\gamma - \gamma_1) + 2\epsilon(\sin \gamma - \sin \gamma_1) + (\epsilon^2/4)(\sin 2\gamma - \sin 2\gamma_1)}{(1 - \epsilon^2)^{5/2}}$$

where the angle γ is defined implicitly as

$$\cos \theta = \frac{\cos \gamma + \epsilon}{1 + \epsilon \cos \gamma}$$

and γ is taken to have the same sign as θ. These integrals were first evaluated by Arnold Sommerfeld in 1904. If cavitation occurs, Q must be chosen such that $dP/d\theta$ and P goes to zero at the same value of θ, which we denote as θ_c, the cavitation angle. Then, for $\theta > \theta_c$, the pressure is assumed to be zero. In practice, it makes little difference whether the pressure is taken as gauge or absolute for the cavitation calculation, since the difference is usually small compared to the maximum pressure that occurs in the bearing.

Once the pressure is known, the vertical and horizontal components may be integrated around the journal to find the magnitude and direction of the resultant load that the bearing supports. The angle between the load and the line of centers is known as the *attitude angle*. In actuality, the position of the load

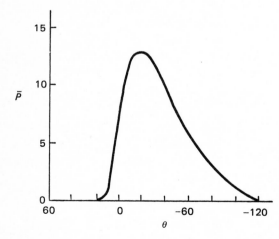

Figure 3-8 Plot of $\bar{P}$ versus θ for 180° partial journal bearing with $\epsilon = 0.8$ and a minimum film thickness 120° from the inlet. The value of θ_c is approximately 20°. Here, $\bar{P} = Pc^2/\mu Ua$ and $\bar{Q} = Q/Uc$.

determines the location of the point of minimum film thickness, but for purposes of calculation it is usually simpler to assume the point of minimum film thickness and ϵ to be known and calculate the load vector.

As an example, the pressure is shown on a dimensionless plot in Fig. 3-8 for a $180°$ bearing with $\theta_1 = -2\pi/3(-120°)$ and $\theta_2 = \pi/3(60°)$. The point of minimum film thickness (the origin) is assumed known and the eccentricity ratio ϵ is taken as 0.8. We define $\bar{P} = Pc^2/\mu Ua$ and $\bar{Q} = Q/Uc$. Cavitation occurs at $\theta \approx 20°$ and $\bar{Q} = 0.122$. The calculation of the load components is then straightforward but is not carried out here.

For the design engineer, extensive charts and tables that are the result of computer calculations are available for infinite- and finite-width journal bearings. These data are based on the theory we have outlined above and are in fact one of the most useful applications of viscous flow theory.

PROBLEMS

3-1 Solve for the pressure distribution and load in a hydrostatic thrust bearing using air as a lubricant. Neglect inertia and assume the flow to be isothermal.

Answer:

$$\frac{P_i^2 - P^2}{P_i^2 - P_o^2} = \frac{\ln(r/a)}{\ln(b/a)}$$

3-2 Derive in detail the expression for the friction coefficient on a hydrodynamic step bearing. Find the total shear force on the bearing surface and check by calculating the total drag force on the slider. Hint: Remember that the pressure P_m acting on the vertical step contributes to the drag.

3-3 Do the same for the inclined slider bearing.

3-4 Liquid metals are often used as lubricants in certain high-temperature installations. Liquid metals behave like Newtonian liquids.

Sometimes it is possible to increase the performance (that is, decrease the necessary flow rate for a given load) by placing the bearing in a magnetic field.

In particular, consider the ordinary hydrostatic thrust bearing in a magnetic field. A detailed analysis shows that the effect of applying a magnetic field axially is merely to add a term (a body force) in the radial equation of motion. This term is a retarding force and may be represented by a constant multiplied by the radial velocity. In units of force per unit volume, the radial body force is then Ku in the negative r direction, where K is a constant that depends on the magnetic field and the properties of the liquid metal. See Fig. 3-9.

(*a*) Write the equation of motion including this body force term (neglect inertia effects).

(*b*) Solve for $u(r, z)$.

(*c*) Solve for the pressure as a function of r and find the load capacity of the bearing.

Answer:

$$u = \frac{(P_i - P_o)}{r\mu K \ln(b/a)} \left[1 - \frac{\cosh \sqrt{K}z}{\cosh \sqrt{K}(h/2)} \right]$$

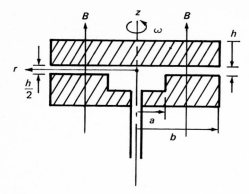

Figure 3-9

where the origin for z is taken as shown in Fig. 3-9. The pressure profile is the same as for the zero-body force solution given in the text.

3-5 Two horizontal parallel round disks of radius a are immersed in a bath of oil. The oil layer is thin, so that the hydrostatic pressure effects in the oil are negligible.

The disks are held at a distance h apart (where $h \ll a$), and suddenly a load W is applied vertically to the top disk and it begins to move toward the lower disk. See Fig. 3-10. This is a model of a squeeze film dynamic problem in bearing analysis.

Your problem is: Find the velocity of the top disk as a function of time. Hint: Assume that motion is slow so that instantaneously the radial flow between the disks is just as for the steady constant h bearing.

3-6 Derive an expression for the pressure at any point in the viscous liquid between two closely spaced parallel circular disks, one of which is fixed and the other approaching with a constant velocity V, as shown in Fig. 3-11.

This problem is similar to Problem 3-5, and similar assumptions may be made.

Further, find the pressure distribution between the disks when the disks are moved apart with a constant velocity V. If the disks are moved too rapidly, the fluid will cavitate. Find the critical value of V such that cavitation occurs if it is exceeded.

3-7 Referring to the analysis (in the text) of the hydrostatic thrust bearing including inertia effects, discuss the possibility of cavitation in a liquid lubricated bearing. (The centrifugal effects can actually cause the fluid film to rupture.)

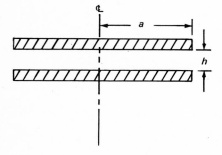

Figure 3-10

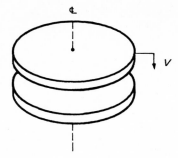

Figure 3-11

What is the criterion for the onset of cavitation? Where does cavitation begin? *Answer:* Cavitation begins at the outer edge if a critical speed given by

$$\Omega^2 \equiv \frac{3\rho a^2 \omega^2}{20(P_i - P_o)} = \frac{1}{2(b/a)^2 \ln(b/a) - (b/a)^2 + 1}$$

is exceeded.

3-8 The design of high-speed rotating oil seals is an art and not completely understood. Consider a simple graphite seal. It consists of a stationary (spring-mounted) graphite "nose

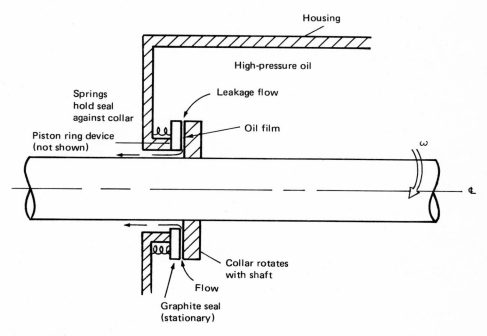

Figure 3-12

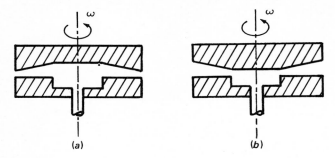

Figure 3-13

piece" in the shape of a flat annular ring (like a washer), which fits flat against a collar attached to the rotating shaft.

In practice, a thin film of the order of 10^{-6} in. of oil is observed between the graphite and collar, although negligible leakage may occur. Various theories have been advanced to explain why a film can be maintained without significant oil leakage.

Your job is to make a preliminary study of what leakage one would expect from simple theory. The configuration is shown schematically in Fig. 3-12. The high-pressure oil is confined on the outside and can leak inward. This is usually the way oil seals are made.

3-9 A hydrostatic thrust bearing has a coning cross section as shown in Fig. 3-13. This coning usually developes naturally from wear due to thermal expansion and stress deformation effects.

For (a) converging and (b) diverging coning, find the pressure distribution in the bearing. Sketch the curves and compare to parallel surface operation.

Now, suppose that one plate is slanted slightly due to misalignment or run-out. Discuss the hydrostatic stability. What would happen if a parallel-plate bearing becomes slightly tilted due to some disturbance in load, etc.?

Since the width $(b - a)$ may be fairly large compared to the separation on a bearing, the actual coning angle may be small. Are the effects still important, or must the angle of coning be large in order for the effects to be significant? Discuss.

On a face seal the recess is not load-supporting and the width (lip) is fairly small, perhaps such that $(b - a) \ll b$. Are the stability effects more pronounced in such a seal as compared to a bearing?

BIBLIOGRAPHY

General

Pincus, O., and B. Sternlicht: "Theory of Hydrodynamic Lubrication," McGraw-Hill Book Company, New York, 1961.

Shaw, M. C., and E. F. Macks: "Analysis and Lubrication of Bearings," McGraw-Hill Book Company, New York, 1949.

Wilcock, D. F., and E. R. Booser: "Bearing Design and Application," McGraw-Hill Book Company, New York, 1957.

Historical

Sommerfeld, A.: Zur hydrodynamischen Theorie der Schmiermittelreibung, *Z. Angew. Math. Phys.*, vol. 50, pp. 97–155, 1904. (The first analysis of a journal bearing was given in this paper.)

FOUR

ENERGY AND HEAT FLOW

4-1 INTRODUCTION

If the thermodynamic properties of a fluid are independent of temperature, the equation of motion and the continuity equation may be solved together to give the velocity profile and pressure profile (dynamic behavior) throughout the fluid for any particular flow problem. However, if such is not the case and one or more properties depend on temperature, then the energy equation and possibly an equation of state must be solved simultaneously with the equation of motion and the equation of continuity in order to determine the dynamic and thermodynamic behavior.

For example, if the fluid is incompressible and viscosity is assumed not to depend on temperature, we can solve for the flow velocity and pressure profiles using only the equation of motion and continuity. Such was the case when we discussed various incompressible viscous flow problems in the previous chapters. Indeed, we additionally assumed the viscosity to be constant, although we could have obtained solutions if the viscosity dependence on pressure had been considered.

We are confining our study throughout this book to incompressible flow with constant viscosity, and shall generally solve the equation of motion without recourse to the energy equation.

Why, then, introduce the concept of an energy balance and thermodynamic relationship in a fluid? The reason is that even in an incompressible fluid with constant viscosity we must make use of the energy equation to determine the temperature profile in a fluid. The knowledge of the temperature throughout a fluid is essential to the determination of heat transfer through a fluid, and as we shall see, of vital importance in studying the boundary layer and convective heat transfer.

Furthermore, it should be emphasized that whereas we may often determine the dynamic behavior of a fluid without reference to the energy equation, the converse is not generally true. In order to solve the energy equation (in all but a few simple situations), it is necessary to know the velocity distribution throughout the fluid, since the velocity enters explicitly into the energy equation. Even for an incompressible fluid, this latter statement still holds. The rate of viscous dissipation depends on the shear stresses (and hence velocity gradients), and the transport of internal energy by the moving fluid depends on velocity. Both of these effects must enter into an energy balance on a control volume containing fluid.

For an incompressible fluid the procedure for solving a flow problem would be to solve the equation of motion (and continuity) for the velocity and pressure profiles—then afterwards use these velocity profiles to solve the energy equation to determine the temperature profile in the fluid.

We shall derive various forms of the energy equation (for both incompressible and compressible fluids) in this chapter, and later we shall apply the appropriate form of the energy equation to boundary-layer flow and develop the basic concepts of the thermal boundary layer and convective heat transfer.

4-2 THE ENERGY EQUATION IN INCOMPRESSIBLE STEADY FLOW WITH A FULLY DEVELOPED VELOCITY PROFILE

If the velocity profile is fully developed as in Poiseuille and Couette flow so that the velocity is a function of only one variable, then the energy equation in differential form is rather easy to derive and interpret and is a convenient starting point. We shall derive the energy equation for such flow and solve a few problems that will illustrate the basic ideas without the complication of several dimensions.

Let us assume that the viscous flow is fully developed (in velocity) so that u is a function only of y, as we did in Chapter 2. We extract a small element, Fig. 4-1, and write the control volume form of the energy equation (which we

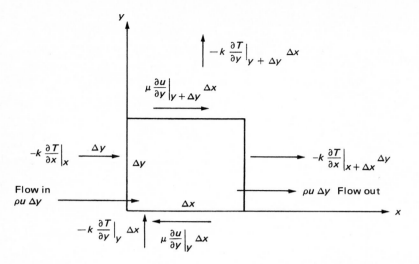

Figure 4-1 Elemental control volume for an energy balance in fully developed (velocity) flow. The temperature is assumed to be a function of x and y, but u, the velocity, is a function of y only.

discussed in Chapter 1) as follows:

$$Q - W_s = \frac{\partial}{\partial t}\int_{\text{vol}} \rho\left(e + \frac{V^2}{2} + \psi\right)d(\text{vol}) + \int_{A} \rho\left(e + \frac{V^2}{2} + \psi + \frac{P}{\rho}\right)\mathbf{V}\cdot d\mathbf{A} \qquad (4\text{-}1)$$

where Q is the rate of heat transfer (by conduction and radiation into the control volume, W_s is the rate at which the fluid in the control volume does mechanical shear (or shaft) work on its surroundings, and e is the specific internal energy. The parameter ψ is the gravitation potential or specific potential energy, which we could write in the familiar form gz, say, if z were the vertical coordinate near the earth's surface. Remember, the force of gravity is defined in terms of ψ as $-\partial\psi/\partial x_i$ per unit mass. Hence if z is the vertical coordinate, the force is simply $-\partial(gz)/\partial z$ or $-g$ (in English units, $lb_f/slug$ and in SI units N/kg).

The terms $\int\rho\,[e + (V^2/2) + \psi]\,\mathbf{V}\cdot d\mathbf{A}$ represent a rate of flux of energy (internal, kinetic, and potential), though the control volume and the term $\int\rho(P/\rho)\mathbf{V}\cdot d\mathbf{A}$ represents the rate at which the pressure forces do work on the surroundings as the fluid flows across the boundary of the control volume. This term together with W_s accounts for the total work rate (reversible and irreversible) by the fluid on its surroundings: work of dilatation, deformation, and translation.

Returning to fully developed flow, we apply Eq. (4-1) to the element of Fig. 4-1. Let us neglect radiation for a while and consider Q to represent conduction only. By Fourier's law, the heat conduction through a unit area may be written as $q_i = -k(\partial T/\partial x_i)$, or in vector form, $\mathbf{q} = -k\,\nabla T$, where $\mathbf{q}$ is defined as the heat flux vector. Combining the heat flow terms over the faces of the elemental cube, we have for the net flow rate into the cube

$$Q = -k\frac{\partial T}{\partial x}\bigg|_x \Delta y + k\frac{\partial T}{\partial x}\bigg|_{x+\Delta x} \Delta y - k\frac{\partial T}{\partial y}\bigg|_y \Delta x + k\frac{\partial T}{\partial y}\bigg|_{y+\Delta y} \Delta x$$

which in the limit (after dividing by $\Delta x\,\Delta y$) as Δx and $\Delta y \to 0$ becomes

$$Q = k\left(\frac{\partial^2 T}{\partial x^2} + \frac{\partial^2 T}{\partial y^2}\right) \tag{4-2}$$

where we have assumed k, the thermal conductivity, to be a constant.

The $(-W_s)$ term can be expressed as

$(-W_s) =$ rate of shear work on the fluid in the volume by the surroundings

$$= \overbrace{\mu\frac{\partial u}{\partial y}}^{\substack{\text{shear stress}\\\text{on top of}\\\text{element, }\sigma_{yx}}} \overbrace{u|_{y+\Delta y}}^{\substack{\text{velocity}\\\text{along}\\\text{top}}} \Delta x - \overbrace{\mu\frac{\partial u}{\partial y}}^{\substack{\text{shear stress}\\\text{on bottom of}\\\text{element, }\sigma_{yx}}} \overbrace{u|_y}^{\substack{\text{velocity}\\\text{along}\\\text{bottom}}} \Delta x$$

which becomes in the elemental limit

$$-W_s = \frac{\partial}{\partial y}(\sigma_{yx}u) = \mu\frac{\partial}{\partial y}\left(\frac{\partial u}{\partial y}u\right) = \mu\frac{\partial^2 u}{\partial y^2}u + \mu\left(\frac{\partial u}{\partial y}\right)^2 \tag{4-3}$$

The first term on the right-hand side, $\mu(\partial^2 u/\partial y^2)u$, is the rate at which the shear stresses do work of translation on the fluid in the element, and the second term, $\mu(\partial u/\partial y)^2$, is the rate at which the shear stresses do irreversible work of deformation, which is a dissipation or friction-work term.

The right-hand side of Eq. (4-1) may be expressed for the elemental control volume as follows (discarding the unsteady terms, dividing by $\Delta x\,\Delta y$ and taking the limit):

$$\frac{\partial[\rho u(e + \psi + P/\rho)]}{\partial x} \tag{4-4}$$

Now, combining (4-2), (4-3), and (4-4) and expanding some of the terms, we have for the energy equation, treating ρ as a constant:

$$\rho u \frac{\partial e}{\partial x} + \rho u \frac{\partial \psi}{\partial x} + u \frac{\partial P}{\partial x} = +k\left(\frac{\partial^2 T}{\partial x^2} + \frac{\partial^2 T}{\partial y^2}\right) + \mu u \frac{\partial^2 u}{\partial y^2} + \mu\left(\frac{\partial u}{\partial y}\right)^2$$

$$-\left(e + \psi + \frac{P}{\rho}\right)\frac{\partial}{\partial x}(\rho u) \tag{4-5}$$

The double-underlined term is zero by continuity for fully developed flow. The single-underlined terms in the above equation form precisely the x component of the equation of motion (in steady, fully developed flow) multiplied by u:

$$u \mid 0 = u \mid -\frac{\partial P}{\partial x} + \mu \frac{\partial^2 u}{\partial y^2} - \rho \frac{\partial \psi}{\partial x}$$

Hence the single-underlined terms in Eq. (4-5) represent work balance among the forces due to pressure gradient, viscous forces, and gravity forces. If the flow were not fully developed or unsteady, it would also be necessary to include the kinetic energy in the balance. This we shall do later in this chapter. We should realize, then, that the original balance, Eq. (4-1), is made up of two separate energy balances, a mechanical work and kinetic energy part, and a thermo- dynamic or first-law balance. These two may be separated as we have just done above. The mechanical balance will always be given by multiplying the appro- priate equation of motion by the velocity. In general, this must be done vectorially by taking the scalar dot product of the velocity $\mathbf{V}$ with the vector equation of motion.

Finally, then, the energy equation we want for steady, incompressible, fully velocity developed flow is

$$\rho u \frac{\partial e}{\partial x} = k \left(\frac{\partial^2 T}{\partial x^2} + \frac{\partial^2 T}{\partial y^2} \right) + \mu \left(\frac{\partial u}{\partial y} \right)^2 \tag{4-6}$$

For a perfect gas de may be written $c_v \, dT$, where c_v is the specific heat at constant volume [not $d(c_v T)$, remember], so that

$$\rho u c_v \frac{\partial T}{\partial x} = k \left(\frac{\partial^2 T}{\partial x^2} + \frac{\partial^2 T}{\partial y^2} \right) + \mu \left(\frac{\partial u}{\partial y} \right)^2 \tag{4-7}$$

However, for some liquids one can assume that c_v and c_p are nearly the same to a good approximation and use (Eq. (4-7) for a liquid.* One important caution, however, arises when we discuss the specific enthalpy h of a liquid. For a perfect gas h is a function of temperature alone and $dh = c_p \, dT$, where c_p is the specific heat at constant pressure. For a liquid this equation is meaningless, and we must

---

*For some liquids this approximation is not very good. As an example, for mercury $c_p/c_v \approx 1.18$. For water, however, $c_p/c_v \approx 1.003$. For liquids c_p is usually measured experimentally and c_p and c_v are then related by the thermodynamic relationship

$$c_p - c_v = \frac{T\beta^2}{\rho} \bigg/ \frac{1}{\rho} \left(\frac{\partial \rho}{\partial P} \right)_T$$

where β is the coefficient of thermal expansion.

always use $h = e + P/\rho$ even though $de \approx c_v\, dT$. Although e is nearly a function of temperature alone for a liquid, h is definitely not.

Before we discuss the general energy equation applicable to a compressible unsteady flow, let us examine a few applications of Eq. (4-7).

4-3 THE THERMAL PROFILE IN COUETTE FLOW

A particularly simple example of the application of Eq. (4-7) is to fully developed, incompressible, steady Couette flow with zero pressure gradient and with the velocity profile given by Eq. (2-13). Consider the top moving plate to be held at a fixed temperature T_1 and the bottom stationary plate at temperature T_0 (Fig. 4-2). Substituting this velocity profile directly into Eq. (4-7), we immediately obtain

$$\rho u\, c_v \frac{\partial T}{\partial x} = k\left(\frac{\partial^2 T}{\partial x^2} + \frac{\partial^2 T}{\partial y^2}\right) + \mu \frac{U^2}{h^2}$$

We assume further that the temperature profile is also fully developed so that T is a function only of y. This assumption is certainly valid if the plates are long enough and the velocity is fully developed. The result is simply

$$\frac{\partial^2 T}{\partial y^2} = -\mu \frac{U^2}{kh^2} \tag{4-8}$$

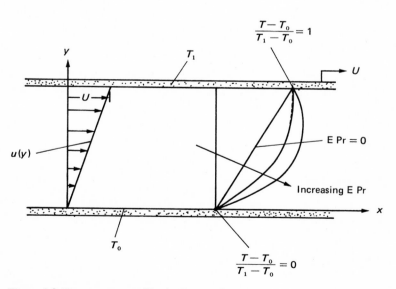

Figure 4-2 Temperature profiles in Couette flow.

Integrating with the boundary conditions $T = T_0$, $y = 0$; $T = T_1$, $y = h$, we obtain

$$\frac{T - T_0}{T_1 - T_0} = \frac{y}{h} + \frac{\mu U^2}{2k(T_1 - T_0)} \left(\frac{y}{h}\right)\left(1 - \frac{y}{h}\right) \tag{4-9}$$

which is a quadratic profile. If there is no dissipation due to viscosity, the profile is simply a straight line between T_1 and T_0. The effect of increasing viscosity is to increase the temperature in the fluid over the straight-line value. Of course, the temperature in the interior of the flow region must always be greater than the straight-line value, since there is a heat source (dissipation) and no heat sink in the fluid.

It is convenient to introduce two dimensionless groups of parameters at this time. Later we shall generate these numbers when we normalize the general energy equation and gain more insight into their significance.

The Prandtl number Pr is defined as

$$\text{Pr} = \frac{\mu c_p}{k} = \frac{\nu}{\alpha}$$

where $\alpha = k/\rho c_p$, the thermal diffusivity. The Prandtl number is near unity for gases, about 0.7 for air. For water Pr is about 7 and for oil of the order of 10^3. The Eckert number E is

$$\text{E} = \frac{U^2}{c_p(\Delta T)}$$

where U is a characteristic velocity of the problem (the velocity of the plate here) and ΔT is a characteristic temperature difference $(T_1 - T_0)$ here.

If we introduce a dimensionless coordinate η as $\eta = y/h$, we can write the temperature profile as

$$\frac{T - T_0}{T_1 - T_0} = \eta + \tfrac{1}{2}\text{E Pr } \eta(1 - \eta) \tag{4-10}$$

Figure 4-2 shows the temperature profile. Only for large values of E Pr is the quadratic term important. For air or water U generally must be extremely large to give rise to a significant effect. In engineering practice the internal viscous effects on temperature are usually insignificant except in high-speed (supersonic) boundary layers or in flow with large Prandtl numbers (such as oil in lubricating films).

Physically, the Prandtl number is a measure of the ratio of the viscous diffusion rate (or diffusion of vorticity) in the fluid to the rate of thermal diffusion. Heat diffuses by conduction at a rate given by the diffusivity constant

α; and similarly, a fluid motion imposed at a fluid boundary will diffuse through a viscous fluid according to the diffusivity (of momentum) ν (the kinematic viscosity).

4-4   THE ADIABATIC WALL TEMPERATURE

If one of the walls in the previous Couette-flow configuration is insulated and allowed to "float" thermally, we might ask what its steady temperature will be. This steady temperature is known as the adiabatic wall temperature and is a useful concept which will be important when we discuss boundary-layer theory and aerodynamic flow.

Referring to Fig. 4-2, we set $T = T_0$ at $y = 0$ and $T = T_{\text{adiabatic wall}}$ (which we denote as T_{ad}) at $y = h$. The temperature is given by Eq. (4-8), but the boundary conditions are now $T = T_0$, $y = 0$; $\partial T/\partial y = 0$, $y = h$. This second condition states that the heat flux to or from the top plate at $y = h$ is zero, since heat flux is proportional to $\partial T/\partial y$ (by Fourier's law). We solve Eq. (4-8) subject to these boundary conditions. Then we can set $y = h$ and solve for T_1, which is the adiabatic wall temperature. The results are

$$T - T_0 = \frac{\mu U^2}{kh^2}\left(yh - \frac{y^2}{2}\right) \tag{4-11}$$

At $y = h$, T is T_{ad}:

$$T_{\text{ad}} - T_0 = \frac{\mu U^2}{2k} \tag{4-12}$$

If we define the Eckert number in terms of $T_{\text{ad}} - T_0$, we have

$$\text{E}_{\text{ad}} = \frac{U^2}{c_p(T_{\text{ad}} - T_0)}$$

and

$$\text{E}_{\text{ad}} \text{ Pr} = 2 \tag{4-13}$$

4-5   THE TEMPERATURE PROFILE IN A ROUND PIPE

A round pipe of radius a is held at uniform constant temperature T_0. If the velocity and temperature are fully developed, the energy equation [cylindrical coordinate version of Eq. (4-7)] becomes

$$k\left(\frac{\partial^2 T}{\partial r^2} + \frac{1}{r}\frac{\partial T}{\partial r}\right) = -\mu\left(\frac{\partial u}{\partial r}\right)^2 \tag{4-14}$$

with boundary conditions $T = T_0$ at $r = a$ and $dT/dr = 0$ at $r = 0$. T and u are both functions of r alone, and

$$u(r) = \frac{1}{4\mu}\frac{dP}{dx}(r^2 - a^2)$$

We have, rewriting (4-14),

$$\frac{1}{r}\frac{d}{dr}\left(r\frac{dT}{dr}\right) = -\frac{\mu}{k}\left(\frac{r}{2\mu}\frac{dP}{dx}\right)^2$$

which integrates directly to

$$T - T_0 = \frac{1}{64k\mu}\left(\frac{dP}{dx}\right)^2(a^4 - r^4) \tag{4-15}$$

4-6 THE GENERAL ENERGY EQUATION

In Section 4-2 we derived a special simplified form of the energy equation. We can now extend this derivation to the complete form of the general energy equation for a fluid. Equation (4-1) may be applied in complete form to an elemental volume of fluid. The results lead to a generalization of Section 4-2 to unsteady compressible flow in three dimensions. Although such a procedure is straightforward and not difficult, it is rather laborious. It is much simpler to convert the terms of the integral energy equation (4-1) into volume integrals and use Gauss' theorem (Section 2-7) to obtain the general differential equation immediately.

Consider each term. Q may be expressed as the flux of heat (by conduction or radiation) into the control volume (Fig. 4-3). In terms of the heat flux vector **q**, the net rate of heat flow *into* the control volume is

$$Q = -\int_A \mathbf{q} \cdot d\mathbf{A} \tag{4-16}$$

The term $(-W_s)$ is the rate at which work is done on the fluid in the control volume by the surroundings, except that work done by the pressure at the boundary. The W_s work term, then, includes all the work done by the stress tensor σ'_{ij} (which includes the shear stresses and any normal stresses other than the isotropic pressure P). Remember that we defined

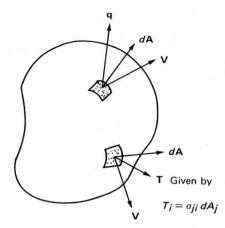

$$T_i = \sigma_{ji}\, dA_j$$

Figure 4-3 A control volume for the energy balance.

$$\sigma_{ij} = \sigma'_{ij} - \delta_{ij}P$$

with $P = -\frac{1}{3}(\sigma_{11} + \sigma_{22} + \sigma_{33})$. The total work rate (deformation plus translation) done by the σ'_{ij} stresses on the fluid in the control volume may be written

$$(-W_s) = \int_A \sigma'_{ij} V_j\, dA_i \tag{4-17}$$

To verify this relationship, consider the product $\sigma'_{ij}\, dA_i$. Physically this is the product of a square matrix and a column vector, the result of which is a vector, called the stress vector T_j, which represents the force (vector) acting on the elemental area dA due to the stress state σ'_{ij}. Remember that the first subscript on σ'_{ij} denotes the face over which the stress acts and represents the coordinate perpendicular to that face. Hence $\sigma'_{ij}\, dA_i$ is physically a vector with three components T_j, two of which lie in the plane of dA and one of which is perpendicular to it. In effect the faces of the cube on which σ'_{ij} is represented are projected onto the face dA. Then $T_j V_j$ is the rate of work done by the stress vector (acting on dA) on the fluid within the control volume.

Gauss' theorem applied to the $(-W_s)$ term is

$$\int_A \sigma'_{ij} V_j\, dA_i = \int_{\text{vol}} \frac{\partial}{\partial x_i} (\sigma'_{ij} V_j)\, d(\text{vol})$$

so that the complete energy equation may be written as a volume integral (interchanging the order of $\partial/\partial t$ and the integral),

$$-\int_{vol} \frac{\partial q_i}{\partial x_i} \, d(vol) + \int_{vol} \frac{\partial}{\partial x_i} (\sigma'_{ij} V_j \, d(vol) - \int_{vol} \frac{\partial}{\partial t} \rho \left(e + \frac{V^2}{2} + \psi \right) d(vol)$$

$$-\int_{vol} \frac{\partial}{\partial x_i} \left[\rho V_i \left(e + \frac{V^2}{2} + \psi + \frac{P}{\rho} \right) \right] d(vol) = 0$$

We may combine all these terms into a single integral which is set to zero. Since the control volume is arbitrary, the integrand itself must be zero, giving the appropriate energy equation in differential form. Expanding out and regrouping, we shall see that the continuity equation will eliminate several terms, and the equation of motion (multiplied by **V**) will appear among the terms representing a mechanical energy balance. By subtracting out this mechanical energy we shall be left with a strictly thermodynamic relationship.

$$\frac{\partial q_i}{\partial x_i} - V_j \frac{\partial \sigma'_{ij}}{\partial x_j} - \sigma'_{ij} \frac{\partial V_j}{\partial x_i} + \left(e + \frac{V^2}{2} + \psi \right) \frac{\partial \rho}{\partial t} + \rho \frac{\partial e}{\partial t} + \rho \frac{\partial (V^2/2)}{\partial t} + \rho \frac{\partial \psi}{\partial t}$$

$$+ \rho V_i \frac{\partial e}{\partial x_i} + \rho V_i \frac{\partial}{\partial x_i} \left(\frac{V^2}{2} \right) + \rho V_i \frac{\partial \psi}{\partial x_i} + V_i \frac{\partial P}{\partial x_i} + \left(e + \frac{V^2}{2} + \psi \right) \frac{\partial (\rho V_i)}{\partial x_i}$$

$$+ P \frac{\partial V_i}{\partial x_i} = 0 \quad (4\text{-}18)$$

The terms with a single underline go out by continuity, and the double-underlined terms are precisely the equation of motion multiplied by the vector **V**. The term $\partial \psi / \partial t$ is zero, since we assume a constant potential with time. We are left, then, with the final form of the energy equation,

$$\rho \frac{De}{Dt} = \rho \left(\frac{\partial e}{\partial t} + V_i \frac{\partial e}{\partial x_i} \right) = -P \frac{\partial V_i}{\partial x_i} - \frac{\partial q_i}{\partial x_i} + \sigma'_{ij} \frac{\partial V_j}{\partial x_i} \quad (4\text{-}19)$$

where D/Dt is the material derivative. The term $\sigma'_{ij}(\partial V_j/\partial x_i)$ is the dissipation rate due to viscous shear, and its functional form in terms of strain rate is known as the *dissipation function*, usually written as Φ. Equation (4-19) may be written in general vector form as

$$\rho \frac{De}{Dt} = -P \nabla \cdot \mathbf{V} - \nabla \cdot \mathbf{q} + \Phi \quad (4\text{-}20)$$

Now, if we assume that **q** is due to conduction given by Fourier's law $\mathbf{q} = -k \, \nabla T$, then $-\nabla \cdot \mathbf{q} = \nabla \cdot (k \, \nabla T)$ or simply $+ k \, \nabla^2 T$ if k is assumed constant and

$$\rho \frac{De}{Dt} = -P \, \nabla \cdot \mathbf{V} + k \, \nabla^2 T + \Phi \tag{4-21}$$

Equation (4-21) may be expressed in several forms. For a perfect gas, e is a function only of temperature and from the definitions of specific heats,

$$c_v = \frac{\partial e}{\partial T}\bigg|_v$$

and if e is a function only of temperature, we see that $de = c_v \, dT$ and Eq. (4-21) becomes

$$\rho c_v \frac{DT}{Dt} = -P \, \nabla \cdot \mathbf{V} + k \, \nabla^2 T + \Phi \tag{4-22}$$

If the specific enthalpy $h = e + P/\rho$ is introduced, Eq. (4-22) takes the form

$$\rho \frac{Dh}{Dt} = \frac{DP}{Dt} + k \, \nabla^2 T + \Phi \tag{4-23}$$

and for a perfect gas $[c_p = (\partial h/\partial T|_p)]$,

$$\rho c_p \frac{DT}{Dt} = \frac{DP}{Dt} + k \, \nabla^2 T + \Phi \tag{4-24}$$

However, it should be emphasized that Eq. (4-24) is valid only for a perfect gas, and *not* a liquid. Equation (4-23) is valid for a liquid, but is not especially convenient because h is a function of both pressure and temperature. For a liquid, Eq. (4-21) is valid and more useful. Indeed, the form (4-22) is usually used with c_v replaced by simply c, the heat capacity of the liquid. The difference between c_v and c_p is not usually significant for a liquid in normal engineering applications.

4-7 THE DISSIPATION FUNCTION

The viscous dissipation Φ has been written in Cartesian tensor form as $\sigma'_{ij}(\partial V_j/\partial x_i)$. This form is valid for any type of fluid, but must be expressed in terms of the tensor strain rate in order to be useful. Under the assumption of a Newtonian fluid (which leads to the Navier-Stokes equation of motion), the dissipation function may be written explicitly in Cartesian coordinates as

$$\Phi = 2\mu \left[\left(\frac{\partial u}{\partial x} \right)^2 + \left(\frac{\partial v}{\partial y} \right)^2 + \left(\frac{\partial w}{\partial z} \right)^2 + \frac{1}{2} \left(\frac{\partial u}{\partial y} + \frac{\partial v}{\partial x} \right)^2 + \frac{1}{2} \left(\frac{\partial v}{\partial z} + \frac{\partial w}{\partial y} \right)^2 + \frac{1}{2} \left(\frac{\partial w}{\partial x} + \frac{\partial u}{\partial z} \right)^2 \right]$$

$$+ \lambda \left(\frac{\partial u}{\partial x} + \frac{\partial v}{\partial y} + \frac{\partial w}{\partial z} \right)^2 \tag{4-25}$$

The last term, involving the second coefficinet of viscosity, will be zero for an incompressible fluid and will generally be small even for a compressible fluid. It is seldom that many of the terms in the remainder of the expression are important, and in most situations of practical interest only a few terms will be of importance. For example, in Couette flow, remember, the only significant term was $\mu(\partial u/\partial y)^2$.

Various forms of Φ in other coordinate systems are listed in the Appendix for reference.

4-8 RELATIONSHIP OF THE ENERGY EQUATION TO BERNOULLI'S EQUATION

In Section 2-13 we discussed the integration of the equation of motion along a streamline to give a first integral of motion known as Bernoulli's equation. Often this integral is confused with the energy equation just developed. Obviously, the final form of the energy equation (4-21) does not contain a kinetic energy term and cannot possibly give Bernoulli's equation. However, if we go back to the general integral form, Eq. (4-1), kinetic energy is included. If Eq. (4-1) is integrated along a streamline, terms that constitute the Bernoulli equation will appear together with other terms. This is to be expected since, as we have pointed out, Eq. (4-1) contains not only information about thermal energy and thermodynamic information, but information about kinetic and potential energy (a work balance) as well. This latter momentum or work balance is independent of the thermal relationships, as was illustrated when we subtracted the equation of motion out of the differential form of the energy equation.

In fact, we could use Bernoulli's equation to subtract the momentum balance out of the integral form of the energy equation.

For example, suppose that we integrate Eq. (4-1) between two points along a streamtube or streamline or between two stations in a pipe with mass flow rate m.* For simplicity we assume steady flow (but this is not necessary),

$$q - w_s = \int_1^2 d\left(\frac{P}{\rho}\right) + \int_1^2 de + \frac{V_2^2 - V_1^2}{2} + \psi_2 - \psi_1$$

or expanding,

---

*A streamtube is composed of streamlines. The flow in a tube of unit cross-sectional area is ρV, a constant along the tube in steady flow. The same equations are valid for flow in a pipe of cross-sectional area A. Then $q = Q/m$ and $w_s = W_s/m$, where m, the mass rate of flow, is ρVA.

$$q - w_s = \left(\frac{P_2}{\rho_2} + e_2 + \frac{V_2^2}{2} + \psi_2\right) - \left(\frac{P_1}{\rho_1} + e_1 + \frac{V_1^2}{2} + \psi_1\right) \qquad (4\text{-}26)$$

q and w_s are now Q/m and W_s/m, respectively, that is, on the basis of a unit mass of flowing fluid. Equation (4-26) is often used in pipe flow and is valid for general viscous compressible, but steady flow.

Equation (4-26) may be combined in various ways with the dynamic equations (2-85), (2-86), (2-87) or (2-88) to give a purely thermodynamic relationship. For example, for steady inviscid compressible flow, subtracting (2-85) from (4-26) yields

$$q = e_2 - e_1 + \int_1^2 Pd\left(\frac{1}{\rho}\right) \qquad (4\text{-}27)$$

which will be recognized as the first law of thermodynamics for a system of unit mass doing reversible work $\int_1^2 Pd(1/\rho)$ on its surroundings. [Here we have made use of the relationship of $dh = d(P/\rho) + de$.] Alternatively, one could begin with this system form of the first law of thermodynamics and combine it with Eq. (4-26) to give a momentum balance given by Eq. (2-85).

If the flow is assumed to be viscous, steady, and incompressible, Eq. (4-26) may be written as

$$\left(\frac{V_2^2 - V_1^2}{2}\right) + (\psi_2 - \psi_1) + \frac{P_2 - P_1}{\rho} = -w_s + q - (e_2 - e_1) \qquad (4\text{-}28)$$

which is still a total energy equation. By comparing this equation to Eq. (2-87), it is seen that we may lump the terms $[-q + (e_2 - e_1)]$ together. Physically this is a loss term which is due to friction (e.g., pipe friction). It is customary in hydraulic engineering to write Eq. (4-28) with this loss term expressed in feet of flowing fluid as

$$\left(\frac{V_2^2 - V_1^2}{2}\right) + (\psi_2 - \psi_1) + \frac{P_2 - P_1}{\rho} + w_s + gH_L = 0 \qquad (4\text{-}29)$$

where $gH_L = (e_2 - e_1) - q$. As we mentioned in Section 2-5, a dimensionless head loss for pipe flow known as a friction factor f is defined as

$$f = H_L \frac{2gD}{L\bar{V}^2}$$

where D is the pipe diameter, L is its length, and $\bar{V}$ is the mean velocity in the pipe. Generally, f is a function of Reynolds number for laminar and turbulent flow, and Eq. (4-29) is used in the laminar and turbulent regions of pipe flow.

w_s is the mechanical work rate done by unit mass of the flowing fluid (turbine work), and a negative value for w_s indicates work done on the fluid (pump work).

For most liquids the change in internal energy may be approximated closely as $\Delta u = c_v \, \Delta T$, allowing a simple estimation of the temperature rise due to frictional effects. [See the footnote following Eq. (4-7).]

It is important to realize that Eq. (4-28) was derived from the energy equation, but only by comparing it with the general Bernoulli equation (2-87) could we identify the head loss term. Equating the head loss to $-q + (e_2 - e_1)$ effectively removes the only thermodynamic information, and Eq. (4-29) becomes identical to Eq. (2-87), an equation that comes strictly from the equation of motion and does not involve the first law of thermodynamics.

One must remember, in fluid mechanics, that an equation involving kinetic energy can always be derived from the equation of motion, and an independent energy equation must contain thermodynamic information, that is, it must contain information relating thermodynamic properties.

PROBLEMS

4-1 Consider the hydrostatic thrust bearing shown in Fig. 4-4. The bearing rotates with speed ω, and the angular velocity of the fluid is generally greater than the radial velocity. For the purposes of this problem, assume that the recess region is not present and that there is no radial velocity. In other words, consider two flat parallel disks, one fixed, one rotating, with a liquid between them. (The whole system might be immersed in a bath to prevent leakage, but this part of the problem is irrelevant.)

If both plates are held at temperature T_0, find the temperature throughout the liquid as

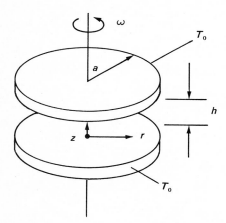

Figure 4-4

a function of r and z. Hint: What is the proper conduction expression? Are all of the terms in that expression important?

4-2 For Problem 2-3 (Chapter 2), calculate the viscous dissipation in the fluid film explicitly. Make a complete energy balance of the viscous pump and verify that the difference between the power in and power out is equal to the total viscous dissipation rate.

4-3 Find the temperature profile in the fluid flowing in an annular pipe as shown in Fig. 4-5. Assume the flow to be steady, laminar, incompressible, and fully developed in velocity and temperature. The inner and outer cylinders are held at temperatures T_0 and T_1, respectively.

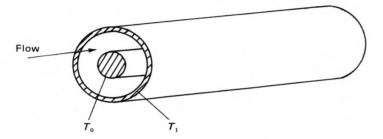

Figure 4-5

4-4 An incompressible fluid is contained between two concentric cylinders that rotate. For steady laminar incompressible flow, find the temperature distribution in the fluid if both cylinders are held at temperature T_0.

4-5 Show that the appropriate energy equation for adiabatic flow in a hydrostatic thrust bearing is as follows, assuming the temperature and properties to be constant across the film. That is, assume that the temperature varies with radial distance only.

For an incompressible fluid,

$$Mc_v \frac{dT}{dr} = 2\pi r \left[\frac{\omega^2 \mu r^2}{h} + \frac{h^3}{12\mu}\left(\frac{dP}{dr}\right)^2 \right] + 2\pi h k \frac{d}{dr}\left(r\frac{dT}{dr}\right)$$

and for a compressible fluid,

$$Mc_p \frac{dT}{dr} = \frac{2\pi r^3 \omega^2 \mu}{h} + 2\pi h k \frac{d}{dr}\left(r\frac{dT}{dr}\right)$$

where M is the mass flow rate, h is the spacing, and ω is the angular velocity of the rotor. k is the thermal conductivity of the fluid.

Hint: Begin with

$$\rho c_v \frac{DT}{Dt} = \mu \left[\left(\frac{\partial v}{\partial z}\right)^2 + \left(\frac{\partial u}{\partial z}\right)^2 \right] + k \, \nabla^2 T$$

for an incompressible fluid or

$$\rho c_p \frac{DT}{Dt} = \frac{DP}{Dt} + \mu\left[\left(\frac{\partial v}{\partial z}\right)^2 + \left(\frac{\partial u}{\partial z}\right)^2\right] + k\,\nabla^2 T$$

for a compressible fluid. Insert the velocity profiles and integrate across the fluid film.

4-6 The Alaskan oil pipeline is laid above ground across hundreds of miles of frozen tundra. The pipe is insulated, and the only heating is that produced by the frictional dissipation in the flowing fluid. Make an energy balance and show that there is indeed enough heat generated by the oil itself to keep it at a reasonable temperature level, and determine the thickness of insulation necessary to keep the oil at, say, 100°F if the outside temperature is −60°F. The following data are given:

Initial delivery rate to be 500,000 bbl/day, increasing to 2,000,000 bbl/day after operation is established. The pipe is 4 ft in nominal diameter of $\frac{1}{4}$-in. steel. (One barrel of oil = 42 gal.) Note that the flow is turbulent and an overall balance is necessary rather than an evaluation of the dissipation function, which we have discussed only for laminar viscous flow.

4-7 Consider a pipe of length L. Obtain the steady flow energy equation for the pipe [as given by Eq. (4-28)] by integrating the general differential form [Eq. (4-20)] throughout the fluid. [For simplicity, assume no turbine or pump, so that w_s is zero in Eq. (4-28).] What happens to the dissipation function term? Do the frictional forces along the wall of the pipe do work?

BIBLIOGRAPHY

Rohsenow, W. M., and H. Choi: "Heat, Mass, and Momentum Transfer," Chap. 7, Prentice-Hall, Englewood Cliffs, N.J., 1961.

Sabersky, R. H., A. J. Acosta, and E. G. Hauptmann: "A First Course in Fluid Mechanics," 2nd ed., Macmillan Company, New York, 1971. (See chap. 3 for a discussion of the Bernoulli equation and its relationship to thermodynamics.)

Schlichting, Hermann: "Boundary Layer Theory," 4th ed., pp. 306, 310, McGraw-Hill Book Company, New York, 1960.

THE BOUNDARY LAYER

5-1 THE BOUNDARY–LAYER CONCEPT

During the nineteenth century the science of hydrodynamics was developed to a high degree of mathematical sophistication. However, this mathematical theory was based on a model of the fluid being inviscid. It was thought that since the viscous forces were often small, they could be neglected entirely with few adverse consequences. Unfortunately, the results of this inviscid theory were more often than not in disagreement with experiment.

In order to predict fluid behavior for engineering design (which was at the time concerned mainly with pipe flow), the science of hydraulics based on experiment and empirical formulation was developed simultaneously with the mathematical hydrodynamics. The disparity between the two approaches perplexed and intrigued scientists of the day, despite the fact that the Navier-Stokes equations had been advanced in the first half of the nineteenth century.

In 1905 Prandtl advanced the hypothesis that the effect of viscosity becomes important in the "boundary layer" and is of similar magnitude to convective acceleration terms. This boundary layer is a thin layer of fluid adjacent to a solid body in a flow stream. On the body the velocity (with respect to the body) must be zero, and far enough away from the body the viscosity may be unimportant. Near the body there must exist a shear flow (with velocity gradients), and the effect of viscosity is important. Outside the boundary layer the flow may be modeled as inviscid, and generally will be irrotational and hence "potential." The

Prandtl hypothesis, then, states that, in general, fluid flow fields may be divided into two parts: a thin boundary layer adjacent to a body, and an inviscid flow region (which may be treated by classical hydrodynamics) throughout the remainder of the flow field (Fig. 5-1). In each region of flow certain simplifications may be made that allow at least approximate analytical solutions to be obtained and experimental data to be correlated in a usable manner.

The boundary layer begins to form at the leading edge or "nose" of any body immersed in a fluid stream. The layer is thin, but grows along the body in the direction of the fluid flow. The layer may be only a few thousandths of an inch thick after growing for a few feet along the surface.

The flow in all boundary layers begins as laminar; then, if the body is sufficiently long, the laminar boundary layer goes through a transition region where large-scale eddies are formed and then develops into turbulent flow in a "turbulent boundary layer." All boundary layers will go turbulent downstream if the body surface is long enough (Fig. 5-2).

The transition to turbulence may be correlated to the Reynolds number, Re_x, based on the downstream coordinate x (Fig. 5-2) and free-stream velocity U_0:

$$Re_x = \frac{xU_0}{\nu}$$

When the Reynolds number exceeds a certain critical value, $Re_{x_{cr}}$, the flow

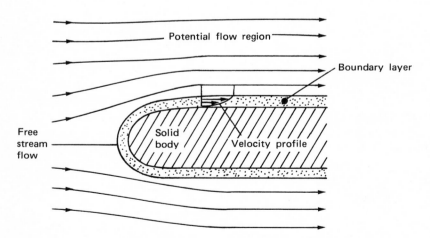

Figure 5-1 The flow region may be divided into two regions, the inviscid region (usually potential flow) and the viscous boundary layer. The fluid velocity is zero on the surface of the object. No slip occurs.

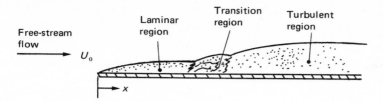

Figure 5-2 Transition from a laminar to a turbulent boundary layer.

undergoes transition to turbulence. The numerical value of $\text{Re}_{x_{cr}}$ depends on the level of turbulence in the free stream and the surface roughness of the body. Increasing either will give a lower value of $\text{Re}_{x_{cr}}$ and promote earlier transition to turbulence.

Several important observations about boundary layers and flow over bodies should be noted before we continue our study in detail.

Transition

A boundary layer always begins laminar and will always develop into a turbulent boundary layer if Re_x exceeds a critical value. This transition occurs regardless of the nature of the free-stream flow, which may be laminar or turbulent.

Skin Friction

The viscous drag on a surface or "skin friction" is simply proportional to the velocity gradient in the boundary layer evaluated at the surface of the body. Hence, once the velocity profile in the boundary layer is known, the shear stress on the wall, τ_w (skin friction) is given by

$$\tau_w = \mu \frac{\partial u}{\partial y}\bigg|_{y=0}$$

referring to Fig. 5-1. Locally the x coordinate is measured along the wall and y is normal to it. As we shall see, the wall shear stress τ_w is greater for a turbulent boundary layer than for a laminar boundary layer.

Laminar Sublayer

The velocity of the fluid at the wall surface or solid boundary must be zero relative to the wall. The Reynolds number (on a local basis) must approach zero near the wall, and one might expect the flow there to be laminar even if the boundary layer itself is turbulent. Indeed this is the case, and a laminar "sub-layer" exists in a turbulent boundary layer. This sublayer is generally much thinner than the turbulent boundary layer itself. However, it is of vital importance in determining the wall shear, where the velocity gradients at the wall is the determining factor. Hence, in a turbulent boundary layer, knowledge of the turbulent velocity profile itself is inadequate to determine the wall shear, which usually must be related to experimental data.

should understand that's in the sublayer

Reynolds Number Effect on Flow over Bodies

If the external flow is such that the Reynolds number based on a characteristic dimension of the object is small (much less than unity), then no definite boundary layer will be formed and the flow, known as "creeping" flow, is analogous to lubrication flow, the inertia being small and negligible to a first approximation. In creeping flow the fluid flows smoothly around the body and viscous effects are felt very far from the body (several characteristic dimensions). Although no wake occurs, the velocity profiles fore and aft are, of course, not symmetrical, because there is drag and momentum loss in the fluid as it flows past the object. The application of creeping flow is important in many engineering problems, but we shall not pursue it in detail here.

The creeping flow around a sphere (Fig. 5-3) has been extensively studied. The classical result of Stokes (1851) gives, for the drag on the sphere (neglecting the inertia of the fluid entirely),

$$D = 6\pi a \mu U$$

where a is the radius of the sphere and U is the free-stream velocity.

For Re < 1, the boundary layer extends out into the entire flow field and has no sharp delineation. It is useful to look qualitatively at the flow field as Re increases to where a definite boundary layer is well defined.

Let us observe the flow around a bluff body, a cylinder as an example here, as the Reynolds number, $\rho D U_0 / \mu$, is increased from much less than unity (creeping flow) through about 100 on to very large numbers of, say, 10^5 or 10^6. In the discussion to follow we shall be concerned primarily with flow corresponding to very large Reynolds numbers, since the transition region from Re $\approx$ 1 to Re $\approx$ 100 is rather complicated. A somewhat indistinct boundary layer becomes well formed at about Re $\approx$ 10. The wake pattern changes from a smooth continuous flow from Re $\lesssim$ 1 to a well-formed standing eddy pattern where streamlines at the rear form closed loops at about Re $\approx$ 6. Then as the Reynolds number increases, the eddy pattern begins to exhibit instability at

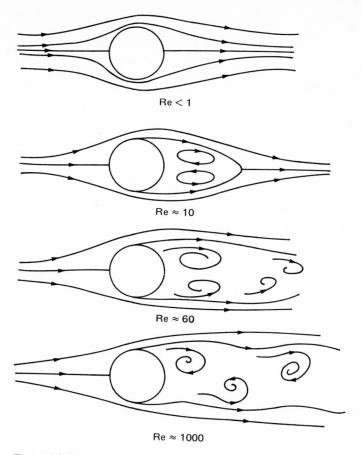

Re < 1

Re ≈ 10

Re ≈ 60

Re ≈ 1000

Figure 5-3 The flow pattern over a cylinder as the Reynolds number increases.

values of about Re ≈ 40. The wake downstream begins to oscillate slowly. At larger Re the oscillation moves upstream, and the two eddies immediately following the cylinder oscillate and appear to form partially and shed alternately from each side of the body. As Re increases to about 100, the forming and shedding becomes quite pronounced and involves the entire vortex. A distinct formation and shedding alternately occurs, and each freshly formed vortex flows downstream, creating a "vortex street" pattern that moves downstream at a velocity somewhat less than U_0.

As Re increases to larger values, the eddies just behind the body become indistinct but the vortex sheet continues to be formed downstream.

Separation and Streamlining

As we just mentioned, as the Reynolds number becomes larger, a distinct wake is formed behind a bluff body, the free-stream flow going around the eddies, which form the wake. The boundary layer actually separates from the surface of the body at the onset of the formation of the wake. Separation is always observed to occur where the free-stream velocity is decreasing. And from Bernoulli's equation we know that the pressure along the surface is then increasing. When boundary-layer separation occurs, it always occurs in the presence of a positive or "adverse" pressure gradient, but such a pressure gradient does not always cause separation. Whether or not a boundary layer actually separates depends on the magnitude of this pressure gradient, and in turbulent flow the history of the boundary layer all along the surface is important.

The prediction of the position where separation occurs is not simple, but it is generally assumed, and experiments confirm, that separation occurs very close to where the wall shear stress goes to zero, which implies that $(\partial u/\partial y)|_{y=0} = 0$. As we shall discuss later, explicit calculation of the zero shear point would involve simultaneous determination of the potential flow field (and boundary layer solution) along with the wake pattern, since one depends on the other. Separation is generally determined experimentally, except in laminar flow where an experimental determination of the pressure distribution may allow explicit calculation of the zero wall shear position.

Hence, if the surface of the body is convex (with respect to the fluid), the boundary layer may become detached from the object and form a region of large-scale turbulence or vorticity near the object. This region is a low-pressure region and develops into a downstream "wake." Separation increases the drag on the object (since a low-pressure region is created downstream) and is generally to be avoided in the design of low-drag bodies. By making the curvature less abrupt, separation of the boundary layers may be avoided and then the object is said to be streamlined (Fig. 5-4).

The Thermal Boundary Layer

The velocity boundary layer has associated with it a "thermal boundary layer" characterized by a temperature profile (analogous to the velocity profile) adjacent to a surface. If a heated body at some given temperature is immersed in a flowing fluid that is at a different temperature, heat will be conducted through the fluid near the surface of the body. This conduction through the moving fluid is generally confined to a thin layer of about the same thickness as the velocity boundary layer. The heat transfer process caused by forcing a fluid over a body

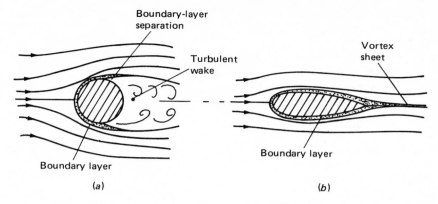

Figure 5-4 Separation and streamlining. (*a*) Separation. A large wake increases drag. (*b*) Streamlining. Only a thin wake (vortex sheet) is present.

is known as forced convection. If the fluid motion is caused by buoyancy effects resulting from the temperature gradients themselves, the process is known as natural convection. Later we shall discuss thermal boundary layers in more detail.

5-2 FLOW PAST OBJECTS AND DRAG

Before we begin a study of the boundary layer, let us first look a bit closer at how the boundary layer enters into actual engineering design.

Flow over objects, or "external flow" as it is called, is an important branch of fluid mechanics and is the very foundation of aerodynamics. As we said earlier, the flow region about an object, airfoil, cylinder, sphere, or any other shape, may be divided into an inviscid or potential flow region and a region near the body where viscous forces are important, the boundary layer.

The behavior of the flow in the inviscid or "potential" region is nearly independent of what happens in the boundary layer so long as separation of the boundary layer does not occur. The boundary layer is thin, and its effect on the potential flow solution is only to increase the apparent size of the body by what is usually a negligible amount. The lift on a body (say, an airfoil) is determined by the pressure forces acting on the body. If the boundary layer does not separate, the net pressure force acting on a body will always be at right angles to the incoming or *free-stream flow* (Fig. 5-5). For an airfoil this is a *lift* force.

However, the net skin friction force due to the boundary layer friction creates a net *drag* force in the direction of the free-stream flow. These forces are nearly uncoupled (if no separation occurs), and the determination of lift and

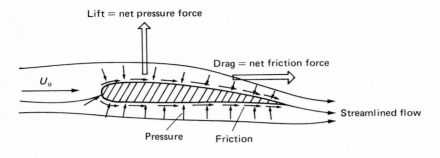

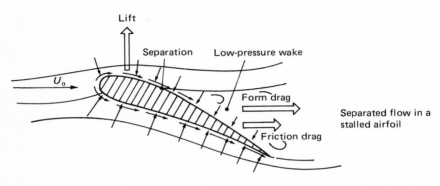

Figure 5-5 The lift is due to net pressure force. If no separation occurs, the drag is due to viscous skin friction. If separation occurs, the drag is the sum of the skin friction and a net pressure force brought about by the low-pressure wake.

drag can often be separated into an aerodynamic or potential-flow problem and a separate boundary-layer problem.

However, if the object is not streamlined, or even if a streamlined object is inclined too much to the free-stream flow, such as the stalled airfoil, boundary-layer separation occurs. Then the net pressure force is not at right angles to the free-stream flow. A low-pressure region in the wake caused by the separation gives rise to a pressure "profile" or "form" drag force in the direction of flow. Usually but not always the form drag is larger than the friction drag.

If separation occurs, the low-pressure wake region behind the body drastically affects the inviscid potential flow region. The wake acts as an extension of the body as far as the inviscid flow region knows, the behavior in the potential flow region becomes coupled to the boundary layer and wake flow, and the entire problem becomes rather complicated, at least from a quantitative point of view.

Drag, then, in general, is made up of two parts—skin friction drag and profile or pressure drag. Under certain conditions one or the other may be larger and dominate the drag behavior depending on whether separation occurs or not. The relative magnitude of the two parts of the drag depends in general on three physical characteristics of the object and fluid: (1) The shape of the body and whether it is streamlined or not. Separation may occur if it is not streamlined. (2) The Reynolds number of the flow based on a characteristic dimension of the object, and (3) the roughness of the surface of the body. How do these three things affect the drag? Let us review some experimental evidence.

From experiments, and later to be confirmed by calculation, we know that the friction drag at a point on a body due to a laminar boundary layer is less than that due to a turbulent boundary layer for a given Reynolds number (that is, for a given free-stream velocity and given kinematic viscosity of the fluid).

For a given object, a laminar boundary layer separates at the same place on the body regardless of the free-stream velocity or kinematic viscosity. A turbulent boundary layer clings to the body better and separates further toward the rear, the more turbulent the layer, hence producing a smaller wake and lower pressure or form drag.

A laminar boundary layer can be made to become turbulent by roughening the body surface. Hence separation may be delayed and form drag reduced by roughening the surface.

To sum up, it would seem that drag might be minimized by either: (1) streamlining and maintaining a laminar boundary layer; or (2) if streamlining is not possible and separation must occur, then a trade-off should be made by reducing form drag at the expense of increased friction drag to achieve a lower overall drag. It is not immediately obvious whether one should try to achieve a laminar or turbulent boundary layer by, say, roughening the surface. As it turns out, it depends on the speed. Figure 5-6 shows a qualitative curve of drag versus speed for a rough and smooth sphere. Up to speed U_1 both spheres have laminar boundary layers and the only difference in drag is a slight effect due to the roughness and small local effects in the boundary layer. The boundary layer on the smooth sphere remains laminar until speed U_2, where it becomes turbulent and the separation point moves toward the rear, drastically reducing the pressure drag. Even though the friction drag becomes greater, the overall drag is reduced since the pressure drag is dominant.

The boundary layer on the roughened sphere, on the other hand, changes to turbulence at a lower speed and hence the drag drops sooner. Eventually, at high speeds the behavior is the same for both spheres except that again (as at very low speeds) the rough sphere has a slightly higher drag.

It is important now to note that the region between U_1 and U_2 is a rather large region covering most of the subsonic aerodynamic region and, for this reason, when separation must occur (as for a ball), a rough surface is usually desirable.

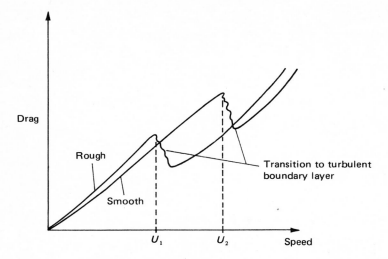

Figure 5-6 The drag-versus-speed curve for flow over two spheres the same size, one roughened and one smooth. After transition to a turbulent boundary layer, the drag drops as the low-pressure wake decreases in size and the form drag drops. Even though the skin friction increases, the overall drag decreases.

Golf balls are roughened. So are baseballs. In fact, for the same driver swing, an undimpled, smooth golf ball will go only about one-fourth as far as a conventional dimpled golf ball.

Force Coefficients

The lift or drag on an object is often determined by experiment. Experiments model the actual situation, and usually smaller "models" are used instead of a full-size "prototype."

From the theory of modeling or "similitude," we know that if the model is made geometrically similar to the prototype, certain relevant important dimensionless parameters may be treated as dependent and independent variables and their functional relationship determined experimentally, independent of the actual size of the model. These relationships also hold for the prototype.

Lift and drag are made dimensionless by dimensional analysis or from an inspection of the governing differential equations. These dimensionless parameters are known as the lift and drag coefficients and are related to the lift and drag as follows:

$$C_f = \frac{D}{\frac{1}{2}\rho U_0^2 A} \qquad C_L = \frac{L}{\frac{1}{2}\rho U_0^2 A}$$

where L and D are lift and drag, respectively, and A is a characteristic area, such as plan form area of a wing, or projected cross-sectional area of a cylinder or sphere. U_0 is the free-stream speed.

For subsonic incompressible flow, C_f and C_L for a given shaped body are mainly functions of the Reynolds number, based on some characteristic length of the body, such as diameter. Our previous statement about the dependence of form and frictional drag on shape, roughness, and Reynolds number can now be stated in a quantitative manner. The general shape and roughness of the surface can be lumped together as the form or geometrical shape of the body, and in general we can write

$$C_f = C_f(\text{Re})$$

$$C_L = C_L(\text{Re})$$

for a given geometrical shape regardless of the actual scale or size of the body.

The coefficients C_f and C_L or, in general, the vector coefficient $\mathbf{C}$, is the important quantity. Plots of $\mathbf{C}$ versus Reynolds number are useful in engineering design and drag data is presented in this manner. However, a plot of C_f versus Re is sometimes not quite as revealing in a qualitative sense as the plot in Fig. 5-6, since the effect of U_0^2 must be taken into account to get the actual drag variation with velocity.

The plot of C_f for a smooth sphere is shown in Fig. 5-7 as an example. The sudden drop in C_f as the boundary layer becomes turbulent is quite noticeable.

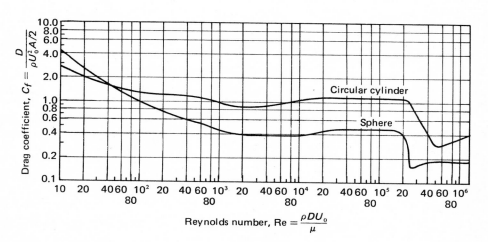

Figure 5-7 Drag coefficient C_f versus Reynolds number $\rho D U_0/\mu$ for a sphere and a cylinder. Data from A. F. Zahm, *NACA Report 253*; and from A. Roshko, *J. Fluid Mech.*, vol. 10, p. 345, 1961.

One other point is important to mention about drag before going on to a detailed study of the boundary layer. At very low Reynolds number flow (Re < 1), where creeping flow takes place no definite boundary layer exists. The inertia forces are negligible and viscous effects extend far out into the flow. No separation occurs, and drag is due solely to skin friction. Hence if the "tail" of the body is streamlined, the drag may actually increase because of the increase in total surface area.

Hence, in summary, the drag on a body immersed in a flowing fluid behaves quite differently depending on the speed (or Reynolds number, actually) of the flow.

At very low speeds (creeping flow), streamlining may do more harm than good. At higher speeds streamlining helps because it prevents separation. If separation occurs because of unavoidable shape design, then roughening the surface helps at higher speeds because the wake is reduced in size and the consequent dominant form drag is reduced. All these factors must be considered in design, and a detailed understanding of the boundary layer is necessary.

5-3 THE BOUNDARY–LAYER EQUATIONS

The basic equations that describe boundary-layer flow of an incompressible fluid are the continuity equation and the Navier-Stokes equation of motion, which we have discussed in Chapter 2. If the fluid is compressible, an equation of state and energy equation will also be needed. We shall confine ourselves here to incompressible flow, which is a good approximation even for gases under low-speed conditions. We shall mention more about compressibility effects later.

As we mentioned, the boundary-layer thickness, δ (Fig. 5-8) is thin compared to its extent in the x direction (direction of flow). Outside the boundary layer, $y > \delta$, the free-stream velocity is $U_0(x)$, which is derived from inviscid flow theory. Since the boundary-layer thickness δ is small compared to physical dimensions of interest in the problem, it is customary to compute $U_0(x)$ at $y = 0$ (on the body) from inviscid theory (usually potential flow theory), neglecting the boundary layer entirely. In other words, the boundary layer has negligible influence on the potential flow unless, of course, separation occurs. Hence $U_0(x)|_{y=0}$ serves as a boundary condition on $u(x)$, the velocity profile in the boundary layer.

The profile $u(x)$ is a continuous function, and in actuality there is no sharp dividing line between the boundary layer and the inviscid flow region. Exact solutions and experiment show that $u(x)$ approaches $U_0(x)$ asymptotically, but it does so rapidly and it is convenient to define $\delta(x)$ as the value of y where $u(y)$

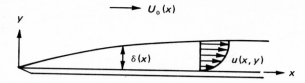

is $0.99U_0$. In some approximate solutions we force u to be exactly U_0 at $y = \delta$, but from an exact point of view $u(\delta) = 0.99U_0$ defines δ, the boundary-layer thickness.

We begin by writing the steady-flow equations of motion in the x and y directions, assuming no variation in the z direction. We consider only a two-dimensional, steady, incompressible boundary layer.

$$x:\ \rho\left(u\,\frac{\partial u}{\partial x} + v\,\frac{\partial u}{\partial y}\right) = -\frac{\partial P}{\partial x} + \mu\left(\frac{\partial^2 u}{\partial x^2} + \frac{\partial^2 u}{\partial y^2}\right)$$

$$y:\ \rho\left(u\,\frac{\partial v}{\partial x} + v\,\frac{\partial v}{\partial y}\right) = -\frac{\partial P}{\partial y} + \mu\left(\frac{\partial^2 v}{\partial x^2} + \frac{\partial^2 v}{\partial y^2}\right)$$

$$(5\text{-}1)$$

Now we make an order-of-magnitude study of the various terms. We assume the range of y to be of order δ and the range of x of order L, where L is some characteristic length in the x direction. Then, from continuity,

$$\frac{\partial u}{\partial x} + \frac{\partial v}{\partial y} = 0$$

We see from order-of-magnitude examination

$$\frac{\partial u}{\partial x} + \frac{\partial v}{\partial y} = 0$$
$$\downarrow \qquad \downarrow$$
$$\frac{u}{L} \qquad \frac{v}{\delta}$$

that $v/u \approx O(\delta/L)$, that is, that v/u is of the order δ/L and hence since $\delta \ll L$, $v \ll u$. Then applying this information to Eq. (5-1), we have

$$x:\ \rho\left(u\,\frac{\partial u}{\partial x} + v\,\frac{\partial u}{\partial y}\right) = -\frac{\partial P}{\partial x} + \mu\left(\frac{\partial^2 u}{\partial x^2} + \frac{\partial^2 u}{\partial y^2}\right)$$
$$\qquad\downarrow \qquad\quad \downarrow \qquad\qquad\qquad \downarrow \qquad \downarrow$$
$$\qquad\frac{u^2}{L} \qquad \frac{vu}{\delta} \qquad\qquad\qquad \frac{u}{L^2} \qquad \frac{u}{\delta^2}$$

$$y: \quad \rho\left(u\frac{\partial v}{\partial x} + v\frac{\partial v}{\partial y}\right) = -\frac{\partial P}{\partial y} + \mu\left(\frac{\partial^2 v}{\partial x^2} + \frac{\partial^2 v}{\partial y^2}\right)$$

$$\downarrow \qquad \downarrow \qquad\qquad\qquad \downarrow \qquad \downarrow$$

$$\frac{uv}{L} \qquad \frac{v^2}{\delta} \qquad\qquad\qquad \frac{v}{L^2} \qquad \frac{v}{\delta^2}$$

In the x equation of motion we can neglect $\partial^2 u/\partial x^2$ compared to $\partial^2 u/\partial y^2$, but we must retain all other terms. However, in the y equation of motion all terms are of at least order v/u less than the corresponding term in the x equation of motion. Hence so must be the $\partial P/\partial y$ term. Consequently, we conclude, to at least order δ/L, that $\partial P/\partial y$ is negligible and the pressure in the boundary layer is a function of x only, not y.

The final result is that the following equations describe the boundary layer and are known as the boundary-layer equations:

$$\rho\left(u\frac{\partial u}{\partial x} + v\frac{\partial u}{\partial y}\right) = -\frac{\partial P}{\partial x} + \mu\frac{\partial^2 u}{\partial y^2}$$

$$\frac{\partial u}{\partial x} + \frac{\partial v}{\partial y} = 0$$

$$(5\text{-}2)$$

The main problem is to find the velocity profile $u(y)$, and hence $\delta(x)$ and the shear stress τ on the wall, given $P(x)$ and $U_0(x)$ from the inviscid solution.

The equations are nonlinear, and exact solutions can only be described in terms of infinite series, which we shall discuss presently.

However, several approximate methods of solutions have been developed and give surprisingly accurate results. These methods are generally based on fitting a polynomial expression for $u(y)$ to an integral equation that takes the place of the differential equations (5-2). The appropriate integral equation may be obtained by either integrating Eqs. (5-2) across the boundary layer or by applying a control volume momentum balance to the boundary layer.

We shall integrate the differential equations from 0 to δ, across the boundary layer.

$$\rho\int_0^\delta \left(u\frac{\partial u}{\partial x} + v\frac{\partial u}{\partial y}\right) dy = -\int_0^\delta \frac{\partial P}{\partial x}\, dy + \mu\int_0^\delta \frac{\partial^2 u}{\partial y^2}\, dy$$

$$= -\delta\frac{\partial P}{\partial x} + \int_0^\delta \frac{\partial}{\partial y}\mu\left(\frac{\partial u}{\partial y}\right) dy$$

$$= -\delta\frac{\partial P}{\partial x} - \tau_w \qquad\qquad (5\text{-}3)$$

where $\tau_w = \mu(\partial u/\partial y)|_{y=0}$ and is the shear stress exerted on the wall by the fluid. The shear stress $\mu(\partial u/\partial y)$ is zero at $y = \delta$. We rewrite the left-hand side of Eq. (5-3) as

$$\rho \int_0^\delta \left[u \frac{\partial u}{\partial x} - u \frac{\partial v}{\partial y} + \frac{\partial}{\partial y}(uv) \right] dy$$

and using continuity this expression becomes

$$\rho \int_0^\delta \left[\frac{\partial}{\partial x}(u^2) + \frac{\partial}{\partial y}(uv) \right] dy$$

Interchanging the order of differentiation and integration since the function must be continuous,

$$\rho \frac{\partial}{\partial x} \int_0^\delta u^2 \, dy + \rho U_0 v|_{y=\delta} = \rho \int_0^\delta \left(u \frac{\partial u}{\partial x} + v \frac{\partial u}{\partial y} \right) dy$$

To find $v|_{y=\delta}$ we use continuity:

$$\frac{\partial}{\partial x} \int_0^\delta u \, dy + \int_0^\delta \frac{\partial v}{\partial y} \, dy = 0$$

$$\frac{\partial}{\partial x} \int_0^\delta u \, dy = -v|_{y=\delta}$$

remembering that $v = 0$ at $y = 0$. Combining, we have finally for the integral equation,

$$\rho \frac{d}{dx} \int_0^\delta (U_0 - u)u \, dy - \rho \frac{dU_0}{dx} \int_0^\delta u \, dy = \delta \frac{dP}{dx} + \tau_w \qquad (5\text{-}4)$$

Often dP/dx is written in terms of U_0 from the equation of motion written just outside the boundary layer along the surface of the body.

$$\frac{1}{\rho} \frac{dP}{dx} + U_0 \frac{dU_0}{dx} = 0$$

We recognize that the first integral of this equation is just the Bernoulli equation,

$$\frac{P}{\rho} + \frac{U_0^2}{2} = \text{constant}$$

The partial derivatives have been changed to ordinary ones since P is not a function of y in the boundary layer, and we have finally

$$\rho \frac{d}{dx} \int_0^\delta (U_0 - u)u \, dy + \rho \frac{dU_0}{dx} \int_0^\delta (U_0 - u) \, dy = \tau_w \qquad (5\text{-}5)$$

It is important to note that Eq. (5-5) is good for laminar or turbulent flow, since no assumptions have been made about u.

Two additional terms are often used in boundary-layer theory, the displacement thickness δ^*, defined as

$$\delta^* = \int_0^\infty \left(1 - \frac{u}{U_0}\right) dy \qquad (5\text{-}6)$$

and the momentum thickness θ, defined as

$$\theta = \int_0^\infty \left(1 - \frac{u}{U_0}\right) \frac{u}{U_0} \, dy \qquad (5\text{-}7)$$

In the approximate methods where a polynomial is used to approximate the velocity profile, the integration must be carried out from zero to δ, and δ^* and θ may be expressed as

$$\delta^* = \int_0^\delta \left(1 - \frac{u}{U_0}\right) dy$$

$$\theta = \int_0^\delta \left(1 - \frac{u}{U_0}\right) \frac{u}{U_0} \, dy$$

In terms of δ^* and θ, Eq. (5-5) becomes

$$\frac{d}{dx}(U_0^2 \theta) + \rho U_0 \frac{dU_0}{dx} \delta^* = \tau_w \qquad (5\text{-}8)$$

The displacement thickness δ^* has a simple physical interpretation. Referring to Fig. 5-9, δ^* is the distance that the wall would have to be displaced outward in order to not change the flow field if the fluid were inviscid. In other words, the flow rate (in the x direction) defect due to boundary layer effects is equal to $\delta^* U_0$.

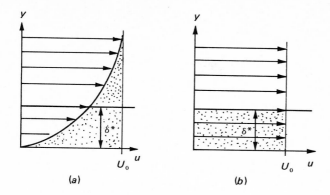

Figure 5-9 Displacement thickness. (*a*) Boundary-layer velocity profile. (*b*) Inviscid velocity profile. The shaded areas in (*a*) and (*b*) are equal.

The momentum thickness θ has the same significance except that it is the momentum flux, not mass flow rate, that is of interest.

The pressure gradient plays an important role, as we have pointed out earlier. On a flat plate where U_0 is a constant (not a function of x), dP/dx is identically zero, but on a curved surface, such as an airfoil, dP/dx is generally a function of x.

We have written the boundary-layer equation in Cartesian coordinates for a flat surface; however, because δ is usually much smaller than any characteristic physical dimension of the body, we can usually neglect curvature of the surface over which the fluid flows and use the equations in Cartesian form.

5-4 THE BOUNDARY LAYER ON A FLAT PLATE

The boundary layer over a flat plate without a pressure gradient is the simplest to study and will allow us to illustrate the equations and methods mentioned above. First an exact solution (in an infinite series form), the Blasius solution, will be discussed, then some approximate solutions will be developed and compared.

Exact Solutions

Exact analytical solutions to the boundary-layer equations usually involve the idea of "similarity." Solutions are "similar" if they are of the same general shape and can be transformed into one another by stretching or shrinking the coordinates, much as a picture drawn on a rubber mat can be stretched in one or more directions. The picture remains "similar" upon stretching the rubber mat.

In boundary-layer theory the velocity profiles $u(y)$ (for most flows) are similar along the x direction (Fig. 5-8). That is, as the boundary layer grows, the velocity profiles at any x position may be made to coincide with the profile at any other x position by a suitable transformation of the y coordinate. The dimensionless velocity (u/U_0) is then a function of only one variable (y/δ).

For a flat plate the suitable similarity transformation suggested by L. Prandtl is as follows. Let

$$\eta = y\sqrt{\frac{U_0}{\nu x}} \tag{5-9}$$

which is equivalent to letting $\eta = Ay/\delta$, where A is some constant, since it will be shown that δ is proportional to $\sqrt{\nu x/U_0}$. In order to reduce the two boundary-layer equations (momentum and continuity) to a single equation, a stream function ψ may be introduced as

$$u = \frac{\partial \psi}{\partial y} \qquad v = -\frac{\partial \psi}{\partial x}$$

which satisfies the continuity equation.* The momentum equation (5-2) (for zero pressure gradient) becomes

$$\frac{\partial \psi}{\partial y} \frac{\partial^2 \psi}{\partial x\, \partial y} - \frac{\partial \psi}{\partial x} \frac{\partial^2 \psi}{\partial y^2} = \nu \frac{\partial^2 \psi}{\partial y^3}$$

Now, using the transformation (5-9), we introduce the following form for the

---

*The continuity equation for an incompressible fluid is

$$\frac{\partial u}{\partial x} + \frac{\partial v}{\partial y} = 0$$

which in terms of the stream function ψ becomes

$$\frac{\partial}{\partial x}\left(\frac{\partial \psi}{\partial y}\right) + \frac{\partial}{\partial y}\left(-\frac{\partial \psi}{\partial x}\right) = 0$$

we see that the equation is always identically satisfied when ψ is defined as indicated above. The stream function ψ is a useful concept only in two-dimensional flow, where ψ has a simple physical meaning. Lines of constant ψ represent streamlines, and the difference between the numerical value of the stream function between two streamlines is equal to the flow rate between those streamlines in incompressible fluid flow.

stream function and corresponding expressions for u and v

$$\psi = \sqrt{\nu U_0 x}\, f(\eta) \qquad \frac{u}{U_0} = f' = \frac{df}{d\eta} \qquad \frac{\partial u}{\partial y} = U_0 \sqrt{\frac{U_0}{\nu x}}\, f'' \qquad v = \frac{1}{2}\sqrt{\frac{\nu U_0}{x}}\, (\eta f' - f)$$

We reduce the equation for ψ to an ordinary but nonlinear differential equation,

$$2\frac{d^3 f}{d\eta^3} + f\frac{d^2 f}{d\eta^2} = 0 \tag{5-10}$$

to be solved for $f(\eta)$. The boundary conditions are

$$f = 0 \qquad \frac{df}{d\eta} = 0 \qquad \left(u = \frac{\partial u}{\partial y} = 0\right) \text{ at } \eta = 0$$

$$\frac{df}{d\eta} = 1 \qquad \left(\frac{u}{U_0} = 1\right) \qquad \text{at } \eta = \infty$$

This equation was solved by H. Blasius (1908) by a power-series expansion. We shall not repeat the details here, but the final result is

$$f = \sum_{n=0}^{\infty} \left(-\frac{1}{2}\right)^n \frac{\alpha^{n+1} C_n}{(3n+2)!} \eta^{3n+2} \tag{5-11}$$

where $\alpha = 0.332$ and the first few values of C_n are

$$C_0 = 1$$
$$C_1 = 1$$
$$C_2 = 11$$
$$C_3 = 375$$
$$C_4 = 27,897$$
$$C_5 = 3,817,137$$

Howarth (1938) calculated numerical values of f, f', and f'', and for convenience we list some values in Table 5-1. Recently, using digital computers, numerical values of f, f', and f'' have been calculated and tabulated for boundary layers with pressure gradients, but we shall not discuss these calculations further. The corresponding velocity profiles are shown in Fig. 5-10.

Several important parameters of the boundary layer may be found immediately from the Blasius solution or Table 5-1. The skin friction along the plate, τ_w, is

Table 5-1 $f(\eta)$ **and Its Derivatives for Flow over a Flat Plate with Zero Pressure Gradients (after Howarth, 1938)**

$\eta = y \sqrt{U_0/\nu x}$	f	$f' = u/U_0$	f''
0	0	0	0.33206
0.2	0.00664	0.06641	0.33199
0.4	0.02656	0.13277	0.33147
0.6	0.05974	0.19894	0.33008
0.8	0.10611	0.26471	0.32739
1.0	0.16557	0.32979	0.32301
1.2	0.23795	0.39378	0.31659
1.4	0.32298	0.45627	0.30787
1.6	0.42032	0.51676	0.29667
1.8	0.52952	0.57477	0.28293
2.0	0.65003	0.62977	0.26675
2.2	0.78120	0.68132	0.24835
2.4	0.92230	0.72899	0.22809
2.6	1.07252	0.77246	0.20646
2.8	1.23099	0.81152	0.18401
3.0	1.39682	0.84605	0.16136
3.2	1.56911	0.87609	0.13913
3.4	1.74696	0.90177	0.11788
3.6	1.92954	0.92333	0.09809
3.8	2.11605	0.94112	0.08013
4.0	2.30576	0.95552	0.06424
4.2	2.49806	0.96696	0.05052
4.4	2.69238	0.97587	0.02897
4.6	2.88826	0.98269	0.02948
4.8	3.08534	0.98779	0.02187
5.0	3.28329	0.99155	0.01591
5.2	3.48189	0.99425	0.01134
5.4	3.68094	0.99616	0.00793
5.6	3.88031	0.99748	0.00543
5.8	4.07990	0.99838	0.00365
6.0	4.27964	0.99898	0.00240
6.2	4.47948	0.99937	0.00155
6.4	4.67938	0.99961	0.00098
6.6	4.87931	0.99977	0.00061
6.8	5.07928	0.99987	0.00037
7.0	5.27926	0.99992	0.00022
7.2	5.47925	0.99996	0.00013
7.4	5.67924	0.99998	0.00007
7.6	5.87924	0.99999	0.00004
7.8	6.07923	1.00000	0.00002
8.0	6.27923	1.00000	0.00001
8.2	6.47923	1.00000	0.00001
8.4	6.67923	1.00000	0.00000
8.6	6.87923	1.00000	0.00000
8.8	7.07923	1.00000	0.00000

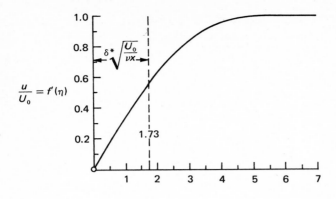

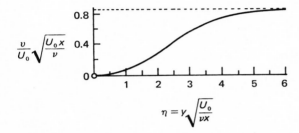

Figure 5-10 The velocity components in the boundary layer on a flat plate, taken from the Blasius solution, 1908.

$$\tau_w = \mu \left. \frac{\partial u}{\partial y} \right|_{y=0}$$

$$= \mu\, U_0 \sqrt{\frac{U_0}{\nu x}}\, f''(0) = \alpha\mu\, U_0 \sqrt{\frac{U_0}{\nu x}}$$

$$\tau_w = 0.332\mu U_0 \sqrt{\frac{U_0}{\nu x}}$$

$$= \frac{0.332\rho U_0^2}{\sqrt{\mathrm{Re}_x}} \tag{5-12}$$

where Re_x is the Reynolds number ($U_0 x/\nu$) based on the distance along the plate x. The skin friction coefficient C_f is defined as

$$C_f = \frac{\tau_w}{\frac{1}{2}\rho U_0^2}$$

so that, from the Blasius solution,

$$\boxed{C_f = 0.664\sqrt{\frac{\nu}{U_0 x}} = \frac{0.664}{\sqrt{\text{Re}_x}}} \tag{5-13}$$

The boundary-layer thickness $\delta(x)$ may be found from Table 5-1. u/U_0 or f' becomes 0.99 at a value of η of approximately 5.0. Hence

$$\delta \approx 5.0\sqrt{\frac{\nu x}{U_0}}$$

or

$$\frac{\delta}{x} \approx \frac{5.0}{\sqrt{\text{Re}_x}} \tag{5-14}$$

where Re_x is the Reynolds number based on distance x, $(U_0 x/\nu)$. And similarly, the displacement thickness and momentum thickness are found to be

$$\delta^* = 1.73\sqrt{\frac{\nu x}{U_0}} \tag{5-15}$$

$$\theta = 0.664\sqrt{\frac{\nu x}{U_0}} \tag{5-16}$$

The total drag on one side of the plate of length l and width b is given by

$$D = b\int_0^l \tau_w \, dx$$

$$= 0.664 b U_0\sqrt{\mu\rho l U_0} = \frac{0.664 b l U_0^2}{\sqrt{\text{Re}_l}}$$

and introducing a mean skin friction τ_{w_m} as

$$D = b l \tau_{w_m} \qquad \tau_{w_m} = \frac{0.664\rho U_0^2}{\sqrt{\text{Re}_l}} = 2\tau_w\big|_{x=l}$$

and a mean drag coefficient C_{f_m} as

$$D = C_{f_m}(\tfrac{1}{2}\rho b l U_0^2) = \tau_{w_m} b l$$

$$C_{f_m} = \frac{\tau_{w_m}}{\tfrac{1}{2}\rho U_0^2}$$

we find

$$C_{f_m} = \frac{1.328}{\sqrt{\text{Re}_l}}$$ (5-17)

where Re_l is $\text{Re}_x|_{x=l} = U_0 l/\nu$, the Reynolds number based on the length of the plate. Remember that these results are good only for laminar flow. Turbulence comes a bit later. Now we shall examine some approximate solutions and see that they give surprisingly good results compared to the Blasius solution.

Some Approximate Solutions

Now we shall solve the flat-plate boundary layer using a polynomial expression for the velocity profile $u(y)$ and Eq. (5-4) (Fig. 5-11).

Following the method of T. von Karman, the idea here is to fit a simple polynomial as closely as possible to $u(y)$, then substitute this expression into Eq. (5-4), integrate over y, and obtain an ordinary differential equation for δ as a function of x. Subsequently, relevant parameters, such as $\delta(x)$, τ_w, C_f, and so on, may be found.

A cubic expression gives quite good results. We assume to begin with that

$$u = a + by + cy^2 + dy^3$$

where a, b, c, and d are constants to be determined. Surprisingly, we need not use the boundary-layer equations to find these constants. We need only use the properties of the velocity profile at $y = 0$ (on the plate) and at $y = \delta$. Note that we use conditions at $y = \delta$ (the outer limit of the layer) and not at $y = \infty$, since we are not expecting an exact solution and want the polynomial to hold only in the region $0 < y < \delta$.

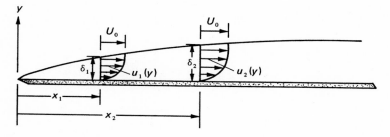

Figure 5-11 "Similarity" of the velocity profile in boundary-layer flow on a flat plate. The function $(u_1/U_0) = f[y/\delta(x_1)]$ is the same as $(u_2/U_0) = f[y/\delta(x_2)]$.

At $y = 0$ we know that the velocity u is zero, and at $y = \delta$ we know that $u = U_0$; we want $\partial u/\partial y|_{y=\delta} = 0$ to ensure a smooth fit. Further, from the momentum equation,

$$\rho\left(u\frac{\partial u}{\partial x} + v\frac{\partial u}{\partial y}\right) = \mu\frac{\partial^2 u}{\partial y^2}$$

we have at $y = 0$ (since $u = v = 0$ there) that $\partial^2 u/\partial y^2 = 0$. These four conditions,

$$
\begin{aligned}
y = 0 \qquad & u = 0 \\
& \frac{\partial^2 u}{\partial y^2} = 0 \\
y = \delta \qquad & u = U_0 \\
& \frac{\partial u}{\partial y} = 0
\end{aligned}
$$
(5-18)

are sufficient to determine a, b, c, and d as

$$a = 0 \qquad b = \frac{3}{2}\frac{U_0}{\delta} \qquad c = 0 \qquad d = -\frac{1}{2}\frac{U_0}{\delta^3} \tag{5-19}$$

Hence the profile is

$$\frac{u}{U_0} = \frac{3}{2}\frac{y}{\delta} - \frac{1}{2}\left(\frac{y}{\delta}\right)^3 \tag{5-20}$$

Before we substitute into the integral expression (5-4), we can immediately extract some useful results from (5-20) itself. The shear stress τ_w can be found in terms of δ as

$$\tau_w = \mu\frac{\partial u}{\partial y}\bigg|_{y=0} = \frac{3}{2}\mu\frac{U_0}{\delta} \tag{5-21}$$

Continuing, we substitute the profile (5-20) into (5-4) to obtain

$$\rho\frac{d}{dx}\int_0^\delta \left\{U_0 - \left[\frac{3}{2}\frac{y}{\delta} - \frac{1}{2}\left(\frac{y}{\delta}\right)^3\right]U_0\right\}\left[\frac{3}{2}\frac{y}{\delta} - \frac{1}{2}\left(\frac{y}{\delta}\right)^3\right]U_0\,dy = \tau_w = \frac{3}{2}\mu\frac{U_0}{\delta}$$

which may be integrated with respect to y to give

$$\rho\frac{d}{dx}\left(\frac{39}{280}U_0^2\delta\right) = \frac{3}{2}\mu\frac{U_0}{\delta}$$

Performing the differentiation, a differential equation for δ is obtained:

$$\frac{39}{280} \rho U_0^2 \frac{d\delta}{dx} = \frac{3}{2} \mu \frac{U_0}{\delta} \qquad (5\text{-}22)$$

Remember that here (on the flat plate) U_0 is a constant. We can separate variables and integrate for $\delta(x)$. The boundary condition is that at the leading edge, $x = 0$, the boundary layer thickness δ is zero. Equation (5-22) may be separated and integrated as

$$\frac{39}{280} \rho U_0^2 \, \delta \, d\delta = \frac{3}{2} \mu U_0 \, dx$$

The integration gives

$$\frac{\delta}{x} = \frac{4.64}{\sqrt{\text{Re}_x}}$$

where $\text{Re}_x = U_0 x / \nu$. The number 4.64 compares favorably with 5.0 of the exact solution [Eq. (5-14)].

Now that $\delta(x)$ is known, skin friction τ_w may be found from Eq. (5-21).

$$\tau_w = \frac{3}{2} \mu \frac{U_0}{\delta} = \frac{0.323 \rho U_0^2}{\sqrt{\text{Re}_x}} \qquad (5\text{-}23)$$

$$C_f = \frac{0.646}{\sqrt{\text{Re}_x}}$$

$$C_{fm} = \frac{1.292}{\sqrt{\text{Re}_x}} \qquad (5\text{-}24)$$

The shear stress τ_w as found by the cubic polynomial approximation is only about 2.7% different from the value found by the exact Blasius solution.

Approximate methods using approximating functions for $u(y)$ are very powerful and important in boundary-layer theory, and they provide simple solutions to otherwise very complicated problems.

Quartic polynomials are often used instead of a cubic as we have just done. However, the improvement in accuracy is very slight. Higher-order polynomials may be easily generated. The coefficients are always found from the boundary conditions at $y = 0$ and $y = \delta$. At $y = \delta$ successively higher derivatives must be zero, and at $y = 0$ the boundary-layer momentum equation may be successively differentiated, then evaluated at $y = 0$ by setting u and v to zero. This procedure gives conditions on higher derivatives of $\partial^2 u / \partial y^2$ at $y = 0$.

The Turbulent Boundary Layer

Thus far we have been discussing the behavior of a "laminar" boundary layer. As we mentioned earlier, however, if the Reynolds number based on distance along the plate, $U_0 x/\nu$, is larger than a critical value, the boundary layer will pass through a transition region and become turbulent. This numerical value of the critical Reynolds number depends on the roughness of the surface and the level of turbulence in the free system. Usually the critical value is of the order of 10^5.

In order to describe the behavior of a turbulent boundary layer in a useful engineering manner, a certain amount of empirical information must be used. Blasius found empirical values for the velocity distribution and wall shear stress. Prandtl proposed for smooth surfaces,

$$\tau_w = 0.0228 \,\rho U_0^2 \left(\frac{\nu}{U_0 \delta}\right)^{1/4} \tag{5-25}$$

and the velocity profile as found by Blasius is[*]

$$\frac{u}{U_0} = \left(\frac{y}{\delta}\right)^{1/7} \tag{5-26}$$

which holds for Reynolds numbers up to about 10^7. Substituting this velocity profile and shear stress into Eq. (5-4) (for zero pressure gradient) and solving for δ, we obtain

$$\frac{\delta}{x} = 0.376 \left(\frac{U_0 x}{\nu}\right)^{-1/5} = 0.376 (\mathrm{Re}_x)^{-1/5} \tag{5-27}$$

where we have assumed that the boundary layer is turbulent from the leading edge, which is not really true—we shall discuss this assumption presently.

Then, the shear stress is found [from Eq. (5-25)] to be

$$\tau_w = 0.0296 \rho U_0^2 \left(\frac{\nu}{U_0 x}\right)^{1/5} = 0.0296 \rho U_0^2 (\mathrm{Re}_x)^{-1/5} \tag{5-28}$$

One difficulty with this calculation is that the assumption has been made

---

[*]Actually, this velocity profile cannot hold in the immediate vicinity of the plate since the calculation of $\mu \, \partial u/\partial y|_{y=0}$ from the velocity profile gives an infinite value of τ_w. Close to the wall, turbulence always dies down and a thin laminar "sublayer" exists. In the sublayer the velocity profile is approximately linear, and outside the sublayer Eq. (5-26) holds. The thickness of the sublayer may be estimated by equating $\mu \, \partial u/\partial x|_{y=0}$ to τ_w from Eq. (5-28) and matching the velocity (assuming a straight line in the sublayer) to that given by Eq. (5-26) with δ from Eq. (5-27) inserted. The result is $\delta_b/\delta = 194/(\mathrm{Re}_x)^{0.7}$, where δ_b is the thickness of the laminar sublayer.

that the turbulent boundary layer began at the leading edge ($x = 0$) with zero thickness ($\delta = 0$). In fact, of course, a laminar boundary layer must precede the turbulent layer until the critical length x_c is reached (Fig. 5-12). The critical value of Re_x, which we denote as Re_{x_c} is then $U_0 x_c / \nu$.

However, the assumption above (due to Prandtl) is fairly good and will be used here although newer, more sophisticated methods have been developed. We assume, then, that when a boundary layer becomes turbulent, the thickness after the transition region at the beginning of the turbulent layer is just what it would be if the turbulent layer had grown from the leading edge.

Hence, in Eqs. (5-27) and (5-28), x is always the value measured from the leading edge, but the equations are applicable only for $x > x_c$.

It is useful to determine the drag coefficient for a plate on which the boundary layer is both laminar and turbulent over different parts of the plate. This will be the situation on any plate of sufficient length. The following calculation illustrates the method that may be used and depends, of course, on the actual value of Re_{x_c}. We carry out the calculation here for $\mathrm{Re}_{x_c} = 10^6$. The drag on a plate of width b and length l is

$$D = b \underbrace{\int_0^{x_c} \tau_w \, dx}_{\text{Laminar}} + b \underbrace{\int_{x_c}^{l} \tau_w \, dx}_{\text{Turbulent}}$$

$$= b \int_0^{x_c} 0.332 \, \rho U_0^2 (\mathrm{Re}_x)^{-1/2} \, dx + b \int_{x_c}^{l} 0.0296 \rho U_0^2 (\mathrm{Re}_x)^{-1/5} \, dx$$

Letting $\mathrm{Re}_x = U_0 x / \nu$ and carrying out the integration, we find that

$$D = \frac{1}{2} \rho U_0^2 bl \left\{ \frac{0.0740}{(\mathrm{Re}_l)^{1/5}} + \frac{1}{\mathrm{Re}_l} \left[\frac{1.328 \mathrm{Re}_{x_c}}{(\mathrm{Re}_{x_c})^{1/2}} - 0.0740 (\mathrm{Re}_{x_c})^{4/5} \right] \right\}$$

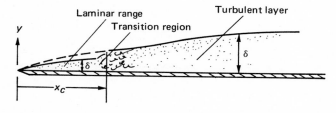

Figure 5-12 The turbulent boundary layer has a thickness at X_c approximately equal to the thickness it would have if it had grown from $x = 0$ as a turbulent boundary layer.

which is true in general for any value of Re_{x_c}. Now, for $\text{Re}_{x_c} = 10^6$, for example, we have

$$D = \frac{1}{2} \rho U_0^2 bl \left[\frac{0.0740}{(\text{Re}_l)^{1/5}} - \frac{3300}{\text{Re}_l} \right] \tag{5-29}$$

or, in terms of C_{f_m} for the plate, where

$$D = \frac{1}{2} \rho U_0^2 bl C_{f_m}$$

we have

$$C_{f_m} = \frac{0.0740}{(\text{Re}_l)^{1/5}} - \frac{3300}{\text{Re}_l} \tag{5-30}$$

as the drag coefficient for a plate with both laminar and turbulent flow and $\text{Re}_{x_c} = 10^6$.

5-5 BOUNDARY LAYERS WITH A PRESSURE GRADIENT

Pressure gradients generally exist along the surface of a body immersed in a flow. The flat plate just discussed had a zero pressure gradient, but if the surface is curved, the free-stream velocity and pressure vary over the surface. Such is the case in flow over airfoils, cylinders, and over any solid body in general. This is an important problem in aerodynamics.

In general, the boundary layer in an "adverse" pressure gradient (one in which $dP/dx > 0$, that is, the pressure increases in the direction of flow) grows more rapidly, and one with $dP/dx < 0$ grows less rapidly than a boundary layer with no pressure gradient.

As we discussed earlier, separation can occur only in the presence of an "adverse" pressure gradient along the surface of the body. Boundary-layer separation can occur naturally in flow over a nonstreamlined body (Fig. 5-13) or from a flat plate if an adverse pressure gradient is imposed. One simple example of separation from a flat surface is separation in a subsonic diffuser (Fig. 5-14a). If the diffuser angle is too great, separation may occur and the resulting pressure recovery is less than it would be for properly designed nonseparating diffusers. The boundary layer may also separate in flow over a sequence of flat surfaces if they form a "salient" angle as shown in Fig. 5-14b.

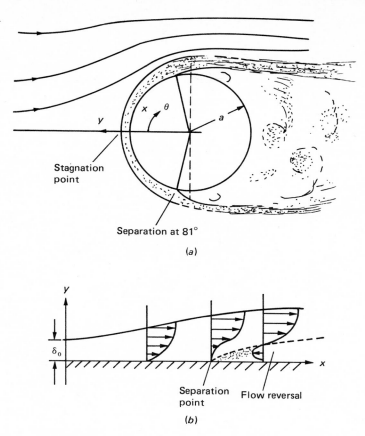

Figure 5-13 A blunt body with stagnation-point boundary-layer thickness and boundary-layer separation. (*a*) Flow over a cylinder. (*b*) Unwrapped boundary layer.

Exact similarity solutions may be effected for some practical flows, but in general numerical solutions are necessary for highly accurate results.* However,

---

*A general similarity solution for the problem of the laminar boundary layer with a variable free-stream velocity U_0 may be obtained if U_0 is of the form

$$U_0 = \beta\left(\frac{x}{c}\right)^m$$

Using the stream function

$$\psi = \sqrt{U_0 \nu x}\, f(\eta)$$

(continued)

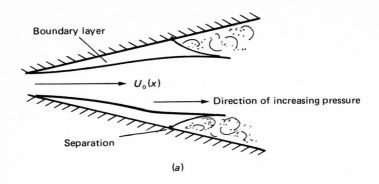

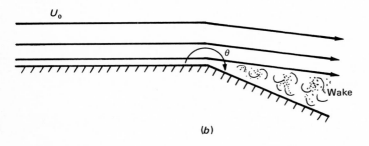

Figure 5-14 (*a*) A subsonic diffuser. If the divergence angle is too great, separation of the boundary layer may occur. (*b*) A "Salient edge" formed by two flat surfaces. The angle θ is salient if greater than $180°$.

$$\eta = \sqrt{\frac{U_0}{\nu x}}\, y$$

The momentum equation takes the form

$$f''' + \frac{1}{2}(m + 1)\, ff'' + m(1 - f'^2) = 0$$

which is known as the equation of Falkner and Skan, who developed solutions for a range of values of m (1937). If U_0 is expanded in a power series of odd power, then

$$U_0 = \sum_{n=0}^{\infty} \beta_{2n+1}\left(\frac{x}{c}\right)^{2n+1}$$

The general solution may be obtained as a series of universal functions, which were tabulated by Howarth (1934) and subsequently improved by others using modern computers. A concise discussion of the general problem is given in the reference by N. Curle listed at the end of this chapter. The tables of these universal functions allow the construction of solutions to a rather broad class of flows with a pressure gradient and, although they will not be discussed here, are of extreme importance in laminar boundary-layer theory.

by using the approximate integral method and polynomial velocity profile, much useful information may be established about the behavior of boundary layers with pressure gradients.

In particular, it is fairly simple to determine the behavior of the boundary layer in the vicinity of the stagnation point on the nose of the body, and some information about separation may be obtained. The solution for the complete boundary layer and details of separation will not be given here.

As for the flat plate, we begin by constructing a polynomial velocity profile. The constants in the polynomial are determined by the same conditions at the outer edge $y = \delta$, but at $y = 0$, the conditions are different. The velocity $u = 0$, but the term $\partial^2 u/\partial y^2$ is now not zero because the equation of motion evaluated at $y = 0$ has the pressure gradient term in it. Successive conditions on the higher derivatives are now modified by the pressure gradient.

However, the quadratic profile remains the same,

$$\frac{u}{U_0} = 2\left(\frac{y}{\delta}\right) - \left(\frac{y}{\delta}\right)^2 \tag{5-31}$$

since the condition on $\partial^2 u/\partial y^2$ is not used. The cubic profile becomes somewhat more complex:

$$\frac{u}{U_0} = \left(\frac{3}{2} + \frac{\lambda}{2}\right)\frac{y}{\delta} - \lambda\left(\frac{y}{\delta}\right)^2 - \left(\frac{1}{2} - \frac{\lambda}{2}\right)\left(\frac{y}{\delta}\right)^3 \tag{5-32}$$

where

$$\lambda = \frac{\delta^2}{2\nu}\frac{dU_0}{dx}$$

The cubic expression is derived as before from

$$u = a + by + cy^2 + dy^3$$

but the conditions used to find the coefficients are now

$$y = \delta \quad \begin{cases} u = U_0 \\[2mm] \dfrac{\partial u}{\partial y} = 0 \end{cases}$$

$$y = 0 \quad \begin{cases} u = 0 \\[2mm] \mu\dfrac{\partial^2 u}{\partial y^2} = \dfrac{dP}{dx} = -\rho U_0\dfrac{dU_0}{dx} \end{cases}$$

The wall shear stress evaluated from the cubic polynomial is

$$\tau_w = \mu\left(\frac{3}{2} + \frac{\lambda}{2}\right)\frac{U_0}{\delta} \tag{5-33}$$

At the point of separation, τ_w is zero, so that at separation $\lambda = -3$, which is a condition that allows the determination of the location of the separation point if $\delta(x)$ is known.[*]

$$\lambda = -3 = \frac{\delta^2}{2\nu}\frac{dU_0}{dx} \tag{5-34}$$

Although dU_0/dx may be known from the external potential flow solution, δ must always be determined from the boundary-layer calculations over the surface.

A major difficulty arises in practice here. The form of dU_0/dx is determined not only by the shape of the body, but also by the point of separation and shape of the wake. Hence, in general, the calculation of boundary-layer flow over a body with separation requires some experimental input usually of the form $P(x)$, since dU_0/dx and $\delta(x)$ depend on each other and the shape of the wake. On streamline bodies such as airfoils, separation does not occur and the calculations for $P(x)$ may be carried out if the shape is known.

The behavior of $U_0(x)$ in the nose or stagnation region is not very sensitive to the separation downstream, and the potential flow solution based on non-separating flow may be used in the vicinity of the stagnation point.

Flow in the Stagnation-Point Region

In general, the potential flow $U_0(x, y)|_{y=0} = U_0(x)$ may be expanded in a power series in x the distance from the stagnation point. In terms of the free-stream velocity U_∞ far from the body, we can write

$$U_0(x) = U_\infty(Ax + Bx^2 + \cdots)$$
$$\frac{dU_0}{dx} = AU_\infty \tag{5-35}$$

[*]It is interesting to note that the quadratic velocity profile cannot be used to predict separation because the value of τ_w is constant. The information about dP/dx is not used in determining the profile.

If a quartic velocity profile is used, the value of λ at separation is -6 instead of -3. The apparent large discrepancy is not as serious as it seems, since the exact point of separation is not extremely sensitive to the value of λ over a broad range of values. The value of -6 is quite close to the exact value, however, and may be used to give good results.

where the coefficients depend on the shape of the body. If we examine the flow near $x = 0$ (the stagnation point), the expression for $U_0(x)$ may be approximated by $U_0 = AU_\infty x$, which allows a simple, explicit solution to the integral form of the equations with a polynomial velocity profile.

The differential equation for $\delta(x)$ (first order) is obtained as for the flat plate by integrating Eq. (5-5) with a suitable velocity profile. The boundary condition on δ is now $d\delta/dx = 0$ at $x = 0$. The value of δ at $x = 0$, δ_0, is called the stagnation-point boundary-layer thickness and is an important parameter in aerodynamic flows.

As an example, we can solve for δ_0 using a quadratic velocity profile. Substituting (5-31) into (5-5) and integrating, we obtain, after a bit of algebra,

$$\frac{x}{15} \frac{d\delta}{dx} = \frac{\nu}{AU_\infty \delta} - \frac{9}{30}\delta \tag{5-36}$$

which may be integrated for $\delta(x)$ but is valid only near the stagnation point. However, the value of δ_0 may be determined immediately since $d\delta/dx = 0$ at $x = 0$, and hence from the equation above the right-hand side must be zero at $x = 0$ and

$$\frac{9\delta}{30} = \frac{\nu}{AU_\infty \delta}$$

and

$$\delta_0 = \sqrt{\frac{30\nu}{9AU_\infty}} \tag{5-37}$$

The analysis may be carried out using a cubic velocity profile, and the result gives $\delta_0 = 2.4 \sqrt{\nu/AU_\infty}$. These approximations are in good agreement with the exact solution, which is $\delta_0 = 2.4 \sqrt{\nu/AU_\infty}$. The cubic approximation is the same as the exact solution to two significant figures.

The Circular Cylinder

The above results may be applied to the circular cylinder (Fig. 5-9) of radius a for which $U_0(x) = 2U_\infty \sin\theta = 2U_\infty \sin(x/a)$ for potential flow around a cylinder with no separation of the boundary layer. Expanding about the origin $x = 0$, we have

$$U_0(x) = 2U_\infty \left[\frac{x}{a} - \left(\frac{x}{a}\right)^3 \frac{1}{3!} + \cdots \right] \tag{5-38}$$

and comparing to the general expansion in the previous section, we see that $A = 2/a$ and the stagnation-point boundary-layer thickness may be obtained explicitly as

$$\delta_0 = 2.4\sqrt{\frac{a\nu}{2U_\infty}} = 1.7\sqrt{\frac{a\nu}{U_\infty}} \tag{5-39}$$

This value depends, of course, on the potential flow solution about the cylinder, and the effect of separation does change the flow and affects the stagnation δ_0 slightly. We have neglected this effect here.

However, the calculation of the boundary layer δ and $u(y)$ around the cylinder cannot be carried out without recourse to experiment, since $U_0(x)$ is dependent on the separation point and shape of the wake. If, however, experiments are carried out to determine $P(x)$ and hence $U_0(x)$, then these values may be used in a numerical calculation around the cylinder to find $\delta(x)$ and the separation point. Such calculations have been made, and they check surprisingly well with the experimental value of the separation point. The angle of separation turns out to be just about 81° measured downstream from the stagnation point. (If no account were taken of the wake and the potential flow solution for nonseparated flow were used, the angle would be about 110°, which is very far from the experimentally observed value.)

Remember, of course, that all of the above statements refer to a laminar boundary layer, and if the flow in the layer is turbulent, the separation point may not be calculated from the basic theory but depends on the level of turbulence and must be found experimentally as a function of the Reynolds number and turbulence level of the flow.

PROBLEMS

5-1 Derive the momentum integral relationship [Eq. (5-5)] by applying the integral form of the momentum equation directly to a control volume, *ABCD*, as shown in Fig. 5-15. Flow occurs across the boundary *B–C*. What direction does the fluid flow across surface *B–C*?

5-2 Using a quadratic velocity profile in the von Karman-Pohlhausen method, find the velocity profile, shear stress, and boundary layer thickness. Compare to the values found using a cubic velocity profile. (Assume that the pressure gradient is zero.)

5-3 Derive Eq. (5-34) in detail. Determine the condition for separation (i.e., the value of λ) using a cubic velocity profile, and verify Eq. (5-34). Show that $\lambda = -3$ at the place where separation occurs.

5-4 Find the appropriate form of the velocity profile for a fourth-degree polynomial for boundary-layer flow with a pressure gradient. Show that the quartic polynomial leads to a value of $\lambda = -6$ at separation.

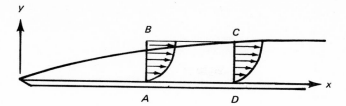

Figure 5-15

5-5 Determine the behavior of the boundary layer in the vicinity of the stagnation point, assuming (a) a quadratic velocity profile, and (b) a cubic velocity profile. In particular, find δ_0 and verify the results for δ_0 given by Eq. (5-37) and the result $\delta_0 = 2.4 \sqrt{\nu/AU_\infty}$ using a cubic velocity profile.

5-6 Derive the suitable boundary conditions to be used to find an appropriate fifth- and sixth-degree polynomial for the velocity profile in the boundary layer. Discuss the effect of a pressure gradient on the boundary conditions.

5-7 A porous flat plate is subjected to suction along its entire length on its underside, which results in a constant velocity (v_0) of the fluid flowing through the plate. The suction velocity is small compared to the constant free-stream velocity U_0. (See Fig. 5-16.)

  At a sufficiently far distance from the leading edge, it is found that the suction stabilizes the boundary layer growth. The boundary-layer thickness and the velocity distribution within the boundary layer are independent of the axial position x.

  In this stabilized region of the boundary layer, find the velocity distribution and the shear stress on the plate.

  Hint: An exact solution may be obtained. Begin with the fundamental equation.

5-8 Verify that with suction at the surface the boundary layer momentum integral equation becomes

$$\tau_w + \rho v_0 U_0 = \rho \left[\frac{d}{dx} \int_0^\delta u(U_0 - u)\, dy + \frac{dU_0}{dx} \int_0^\delta (U_0 - u)\, dy \right]$$

where v_0 is the y component of the fluid velocity at the surface (a negative number).

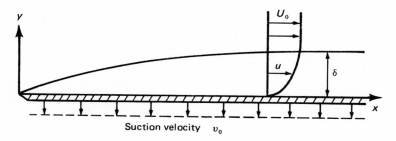

Figure 5-16

5-9 Consider the boundary layer over an object in a flow of known free-stream velocity such that at some point along the surface the boundary layer will be just about to separate. Assume that to prevent this separation from occurring, a suction velocity is impressed on the body surface (from inside) and that suction is applied from the point of incipient separation onto the trailing end of the object. The suction is just sufficient to hold the boundary layer on the verge of separation. How must the suction velocity be related to the free-stream velocity gradient in this region?

Interpret your conclusions in terms of flow over a cylinder or airfoil.

What effect would reversing the sign of v_0 (that is, blowing the boundary layer) have on separation and boundary-layer stabilization?

5-10 Consider air flow at standard conditions over a very wide flat plate of length L. For various vaues of L (1 m, 10 m, 100 m) and various values of U_0 (1 m/s, 10 m/s, 100 m/s) in various combinations, find the boundary-layer thickness for laminar flow at the trailing edge of the plate. Find the total drag on the plate per unit width.

Do the same calculations assuming that the boundary layer is turbulent over the entire plate and compare the results.

Assuming critical Reynolds numbers of 10^5 and 10^6, discuss the location of the transition region and the method of determining the total drag on the plate.

5-11 Repeat Problem 5-10 assuming water for the fluid.

BIBLIOGRAPHY

General

Batchelor, G. K.: "An Introduction to Fluid Dynamics," (secs. 4.9–4.12 and chap. 5), Cambridge University Press, Cambridge, 1967.

Curle, N.: "The Laminar Boundary Layer Equations," Oxford University Press, London, 1962.

Goldstein, S. (ed.): "Modern Developments in Fluid Dynamics," Oxford University Press, London, 1938.

Pai, S. I': "Viscous Flow Theory." Van Nostrand, Reinhold, New York, 1956.

Schlicting, H.: "Boundary Layer Theory," 4th ed., McGraw-Hill Book Company, New York, 1960.

Shapiro, A. H.: "Shape and Flow," Doubleday (Anchor Books), Garden City, N.Y., 1961.

Historical

Blasius, H.: *Z. Math. Phys.*, vol. 56, p. 1, 1908.

Howarth, L.: *Proc. Roy. Soc. Ser. A,* vol. 164, p. 547, 1938.

Pohlhausen, E.: *Z. Math. Mech.*, vol. 1, p. 252, 1921.

Prandtl, L.: Uber Flussigkeitsbewegung bei sehr kleiner Reibung, *Proc. III. Intern. Math. Congress, Heidelberg,* 1904. Reprinted in *NACA Tech. Memo No. 452,* 1928.

von Karman, T.: *Z. Math. Mech.*, vol. 1, p. 233, 1921.

THERMAL EFFECTS
IN BOUNDARY-LAYER FLOW

6-1 THE CONCEPT OF CONVECTION

The study of the transfer of heat is a very important part of engineering. Very few engineering designs are carried out without some consideration of the production and/or transfer of heat.

Heat exchangers are devices that remove heat from one stream of flowing fluid and transfer it to another fluid (usually also in motion). Automobile radiators, residential heating radiators, and boiler water condensers in power plants are a few examples of heat exchangers.

In a heat exchanger the fluid streams (either liquid or gas or a combination of both) are usually separated by a solid interface such as a plate, or one fluid may flow through a pipe or duct while the other flows over it. Basically the heat transfer process consists of heat conduction from a main stream of fluid at some temperature (say, T_0), to a solid surface (plate or wall) at temperature T_w through a thermal *boundary layer*. Since the fluid may move over the wall, the effect of transport of internal energy by the fluid near the wall is important; the overall process of heat transfer between the wall and fluid stream is known as *convection* or convective heat transfer. The process of heat exchange takes place in the thermal boundary layer near the wall. The thermal boundary layer is analogous to the velocity boundary layer.

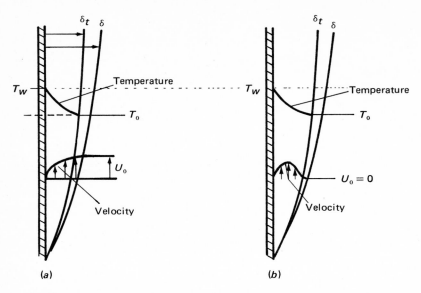

Figure 6-1 Sample velocity and temperature profiles for free and forced convection. The velocity boundary-layer thickness is δ, and the thermal boundary layer thickness is δ_t. (a) Forced convection. (b) Free or natural convection.

There are generally two types of flow that give rise to thermal boundary layers: (1) forced flow caused by blowing or pumping fluid over a plate or wall (or over or through a pipe or duct), which gives rise to *forced convection;* and (2) natural flow, which comes about because of buoyancy effects in the fluid next to the wall. Fluid near a hot vertical wall will become heated and rise relative to the stationary free-stream fluid. Convection that takes place because of this natural flow of fluid is known as *free* or *natural convection.* Sample profiles are shown in Fig. 6-1.

Thermal boundary layers are about the same order of magnitude in thickness as velocity boundary layers for fluids. The thermal layer is thicker for the Prandtl number $\text{Pr} = v/\alpha = \mu c_p/k$ less than unity and thinner for Pr greater than unity. The thermal and velocity boundary layers will be exactly the same in thickness if the Prandtl number is unity. For most gases Pr is slightly less than unity, and for many liquids Pr is of order unity. For oil, however, $\text{Pr} \approx 1000$, and the thermal boundary layer is only about one-tenth as thick as the velocity boundary layer.

We denote the thermal boundary layer thickness as δ_t and define it the same way as the velocity thickness δ. That is, we can let δ_t be the value of y where, $(T_w - T)/(T_w - T_0) = 0.99$.

Now we shall discuss the equations that describe the thermal boundary layer and some of the simple, but very useful solutions. As for the velocity boundary layer, we shall derive the appropriate differential equations and discuss some exact solutions for laminar flow, and then we shall derive an integral form of the equation and discuss some approximate solutions for both laminar and turbulent flow. The aim is to determine the rate of heat transfer between the wall and the fluid free stream.

6-2 THE THERMAL BOUNDARY–LAYER EQUATIONS

We begin by considering the energy equation for a flowing liquid. We must consider liquids and gases separately, but the final result is the same. Relative to a coordinate system fixed to a wall as shown in Fig. 6-2, the general equation for a *perfect gas* (4-24) becomes

$$\rho c_p \left(\frac{\partial T}{\partial t} + u \frac{\partial T}{\partial x} + v \frac{\partial T}{\partial y} + w \frac{\partial T}{\partial z} \right) = \left(\frac{\partial P}{\partial t} + u \frac{\partial P}{\partial x} + v \frac{\partial P}{\partial y} + w \frac{\partial P}{\partial z} \right)$$

$$+ k \left(\frac{\partial^2 T}{\partial x^2} + \frac{\partial^2 T}{\partial y^2} + \frac{\partial^2 T}{\partial z^2} \right) + \Phi \quad (6\text{-}1)$$

where Φ is the viscous dissipation function. In most engineering problems the effect of Φ (effectively an internal heat generation term because of viscous friction) is negligible and becomes important only in very high-speed flow.

Let us consider for the time being any steady flow in which $\partial P/\partial x = 0$ (that is, no pressure gradient). Then we can examine the order of magnitude of the terms in Eq. (6-1). We assume the flow to be two dimensional in the xy plane, with no z flow or variation. An order-of-magnitude study indicates for each term

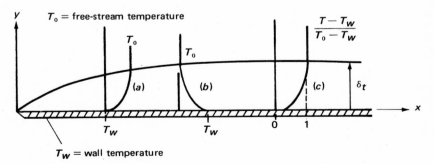

Figure 6-2 Coordinate system for the thermal boundary layer. (*a*) $T_0 > T_w$. (*b*) $T_0 < T_w$. (*c*) Dimensionless plot.

(where δ_t is the thermal boundary-layer thickness, assumed to be roughly of order δ) the following: We divide through by U_0 for convenience,

$$\left(\frac{u}{U_0}\frac{\partial T}{\partial x} + \frac{v}{U_0}\frac{\partial T}{\partial y}\right) = \left(\frac{k}{U_0\rho c_p}\right)\left(\frac{\partial^2 T}{\partial x^2} + \frac{\partial^2 T}{\partial y^2}\right) + \frac{\Phi}{\rho c_p U_0}$$

$$(1)\left(\frac{1}{L}\right) \quad (\delta)\left(\frac{1}{\delta_t}\right) \qquad \delta_t^2 \qquad \left(\frac{1}{L^2}\right) \ \left(\frac{1}{\delta_t^2}\right)$$

(6-2)

Comparing terms, we see that the term $\partial^2 T/\partial x^2$ is negligible compared to $\partial^2 T/\partial y^2$ and that $k/\rho c_p U_0$ must be of the order of δ_t^2. Since, remember, δ^2 is of order $\nu x/U_0$ for laminar flow, we expect that δ_t is of order δ/Pr. We shall derive the exact expression presently.

An order-of-magnitude analysis on Φ indicates that only the term $\mu(\partial u/\partial y)^2$ is important. The final energy equation for the steady thermal boundary layer then becomes

$$\rho c_p\left(u\frac{\partial T}{\partial x} + v\frac{\partial T}{\partial y}\right) = k\frac{\partial^2 T}{\partial y^2} + \mu\left(\frac{\partial u}{\partial y}\right)^2$$

(6-3)

The dissipation term $\mu(\partial u/\partial y)^2$ is generally small for most engineering applications, and we shall neglect it for the time being. Hence we shall be concerned finally with the equation for steady flow with no pressure gradient:

$$\boxed{\rho c_p\left(u\frac{\partial T}{\partial x} + v\frac{\partial T}{\partial y}\right) = k\frac{\partial^2 T}{\partial y^2}}$$

(6-4)

Although Eq. (6-4) was derived for a perfect gas, it is valid for a liquid in steady flow even with a pressure gradient along the flow. We can see this by examining Eq. (4-21). If the flow is incompressible, $\nabla \cdot \mathbf{V} = 0$ and we can let $de \approx c\,dT$, where c is the specific heat for a liquid without designation as to c_v or c_p.* If we say that $c \approx c_p$ (which is true for a liquid), then we arrive directly at Eq. (6-4). It should be remembered that Eq. (4-23), although valid for a liquid, is not very useful because we *cannot* say that $dh = c_p\,dT$ except for a perfect gas.

The boundary conditions for Eq. (6-4) are similar to those for the velocity equation. At $y = 0$, $T = T_w$, and at $y = \infty$, $T = T_0$, and T_w and T_0 may be prescribed functions of x. Of course, u and v must be known before the equation can be solved. For an incompressible fluid the equations are uncoupled, and u and v may be completely determined from the Prandtl boundary-layer equation. If, however, compressibility effects are important, the Prandtl equation and Eq.

*See the footnote following Eq. (4-7).

(6-4), together with an equation of state, would have to be solved together. We shall confine our study here to incompressible fluid flow.

The heat flux rate from the wall to the fluid q_w is

$$q_w = q\,|_{y=0} = -k\,\frac{\partial T}{\partial y}\bigg|_{y=0}$$

The units of q_w are Btu/ft^2 · h or, in SI units, W/m^2. Another boundary condition that is sometimes useful is the specification of $q_w(x)$ with $T_w(x)$ to be determined.

6-3 THE FILM COEFFICIENT AND NUSSELT NUMBER

It is convenient to write the flux of heat q_w in terms of what is known as the *film coefficient h* (measured in Btu/ft^2 · h · °F):

$$q_w = -k\,\frac{\partial T}{\partial y}\bigg|_{y=0} = h(T_w - T_0)$$

Once the solution of Eq. (6-4) is known, q_w can be found as a function of x and hence $h(x)$ can be found in terms of relevant parameters. In engineering heat transfer, the determination of h is the central problem, and convective heat transfer problems are usually solved in terms of the film coefficient h or the Nusselt number.

The dimensionless film coefficient is known as the Nusselt number Nu, and is defined as

$$\mathrm{Nu} = h\!\left(\frac{x}{k}\right) \tag{6-5}$$

We shall presently see physically how the Nusselt number arises in a natural way when the equation is solved.

6-4 EXACT SOLUTIONS TO THE THERMAL BOUNDARY–LAYER EQUATION FOR PR = 1

Once u and v are found, Eq. (6-4) can in principle be solved. For certain simple boundary conditions a "similarity" solution exists, and under the appropriate transformation of variables (6-4) may be reduced to an ordinary linear equation and solved by a series expansion or numerical integration.

By defining similarity parameters as we did for the velocity boundary layer in Chapter 5,

$$\eta = y\sqrt{\frac{U_0}{\nu x}} \qquad \psi = \sqrt{\nu x U_0}\, f(\eta)$$

$$u = U_0 f'(\eta) \qquad v = \frac{1}{2}\sqrt{\frac{\nu U_0}{x}}\,(\eta f' - f)$$

the velocity boundary layer for incompressible flow becomes, as before [Eq. (5-10)],

$$ff'' + 2f''' = 0$$

(where the prime denotes differentiation with respect to η). The energy equation can then be written as

$$\frac{d^2 T}{d\eta^2} + \frac{\text{Pr}}{2} f \frac{dT}{d\eta} = 0 \tag{6-6}$$

and can be solved numerically once f is known.* Such solutions were first arrived at by Pohlhausen (1921), but will not be discussed here. The general results are shown in Fig. 6-3. Rather, we shall now discuss a special solution that is easy to interpret in a physical manner.

An elegant but simple and important solution may be found if the Prandtl number $\text{Pr} = \nu/\alpha$ is unity. That is, if $\nu = \alpha$, the thermal and viscous diffusivities are equal and the velocity and temperature profiles behave in the same way, and $\delta = \delta_t$. We let the wall temperature and free-stream temperature be constants, T_w and T_0, respectively. Then, if a solution to the velocity boundary layer is known, $u(y)$, $\delta(x)$, $\tau_w(x)$, and so on, the thermal boundary layer equation can be established immediately.

Consider the Prandtl equation and energy equation together:

$$\left(u\frac{\partial u}{\partial x} + v\frac{\partial u}{\partial y}\right) = \nu\frac{\partial^2 u}{\partial y^2}$$

(continued)

---

*A general solution for $T(\eta)$ in terms of the function f'' may be easily found, but since f is generally obtained numerically or by series expansion, the explicit general form of $T(\eta)$ cannot be expressed in closed form. The solution for T in terms of f'' is

$$\frac{T - T_w}{T_0 - T_w} = \frac{\displaystyle\int_0^\eta (f'')^{\text{Pr}}\, d\eta}{\displaystyle\int_0^\infty (f'')^{\text{Pr}}\, d\eta}$$

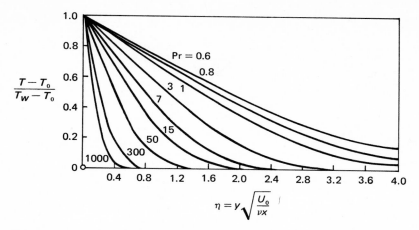

Figure 6-3 The temperature solutions to Eq. (6-6). (After Pohlhausen, 1921.)

$$\left(u\frac{\partial T}{\partial x} + v\frac{\partial T}{\partial y}\right) = \alpha\frac{\partial^2 T}{\partial y^2}$$

These two equations can be made exactly the same as follows. In dimensionless form we can let $u' = u/U_0$ and $T' = (T - T_w)/(T_0 - T_w)$. Then the equations become

$$\left(u\frac{\partial}{\partial x} + v\frac{\partial}{\partial y}\right)u' = \nu\frac{\partial^2 u'}{\partial y^2}$$

$$\left(u\frac{\partial}{\partial x} + v\frac{\partial}{\partial y}\right)T' = \alpha\frac{\partial^2 T'}{\partial y^2}$$

(6-7)

with the boundary conditions $y = 0$, $T' = u' = 0$; and $y = \infty$, $T' = u' = 1$. Hence *if* $\nu = \alpha$ (Pr = 1), the equations are identical in every respect. We expect, then, the solution for (u/U_0) and $(T - T_w)/(T_0 - T_w)$ to be exactly the same functions of x and y. Now, most gases have Prandtl numbers of just about unity (0.7 for air at standard conditions), so such a solution is indeed meaningful.

For a flat plate in longitudinal flow (U_0 parallel to the plate), we know the exact Blasius solution for (u/U_0) and τ_w. We know that

$$\tau_w = \mu\left.\frac{\partial u}{\partial y}\right|_{y=0} = U_0\mu\left.\frac{\partial u'}{\partial y}\right|_{y=0} = \frac{0.332\rho U_0^2}{\sqrt{Re_x}}$$

Hence

$$\left.\frac{\partial u'}{\partial y}\right|_{y=0} = \frac{0.332\rho U_0}{\mu\sqrt{Re_x}}$$

But we also know that

$$\left.\frac{\partial u'}{\partial y}\right|_{y=0} = \left.\frac{\partial T'}{\partial y}\right|_{y=0} = \frac{0.332\rho U_0}{\mu\sqrt{Re_x}} \tag{6-8}$$

and since

$$q_w = -k\left.\frac{\partial T}{\partial y}\right|_{y=0} = -k(T_0 - T_w)\left.\frac{\partial T'}{\partial y}\right|_{y=0}$$

we can combine Eq. (6-8) and the last equation to give q_w explicitly as

$$q_w(x) = \frac{k}{\mu U_0}(T_w - T_0)\tau_w = \frac{k(T_w - T_0)(0.332)\rho U_0}{\mu\sqrt{Re_x}} \tag{6-9}$$

and the film coefficient h may be found from

$$q_w(x) = h(T_w - T_0) = \frac{k(T_w - T_0)(0.332)\rho U_0}{\mu\sqrt{Re_x}}$$

Remembering that

$$Re_x = \frac{\rho x U_0}{\mu} = \frac{x U_0}{\nu}$$

we have

$$\boxed{h_x = \frac{0.332 k U_0}{\nu\sqrt{Re_x}} = \frac{0.332 k \sqrt{U_0}}{\sqrt{x\nu}}} \tag{6-10}$$

We write the local value of h as h_x to emphasize that h depends on x and is a "local" value. At $x = 0$, h goes to infinity. The average value of h over any finite length of plate, however, is finite. The reason for this peculiarity at $x = 0$ is fairly obvious. At the leading edge, where $x = 0$, the oncoming free stream at T_0 is essentially on the plate at T_w (since $\delta_t = 0$ at $x = 0$) and there is, just at $x = 0$, an infinite temperature gradient. But this condition does not persist, and the gradient quickly becomes finite. For any finite area of plate the heat transfer is, of course, finite.

Similarly, the "local" Nusselt number Nu_x may be defined in terms of h_x as

$$\boxed{Nu_x = \frac{(h_x)(x)}{k} = 0.332\sqrt{Re_x}} \tag{6-11}$$

Of particular interest is the total heat flux rate Q from a wall of length l in the x direction (and unit width) and the definition of the average film coefficient $\bar{h}$.

$$Q = \int_0^l q_w \, dx = \int_0^l h(T_w - T_0) \, dx = \bar{h}l(T_w - T_0)$$

Integrating using Eq. (6-11), we obtain

$$\bar{h} = \frac{1}{l} \int_0^l h_x \, dx = \frac{1}{l} \int_0^l \frac{0.332k\sqrt{U_0}}{\sqrt{x\nu}} \, dx = \frac{(2)(0.332)k\sqrt{U_0}}{\sqrt{l\nu}}$$

or

$$\boxed{\bar{h} = 2h_l} \qquad (6\text{-}12)$$

Hence the average value of h_x for the plate of length l is twice the value of h at the end of the plate, $x = l$. The result is analogous to the average value of the wall shear stress on a flat plate.

Similarly, the average Nusselt number $\overline{\text{Nu}}$ may be defined as

$$Q = \int_0^l \text{Nu}_x\left(\frac{k}{x}\right)(T_w - T_0) \, dx = \overline{\text{Nu}}\left(\frac{k}{l}\right)l(T_w - T_0)$$

so that

$$\overline{\text{Nu}} = \frac{1}{l} \int_0^l \text{Nu}_x\left(\frac{kl}{x}\right) dx = \frac{1}{l} \int_0^l 0.332\sqrt{\frac{xU_0}{\nu}}\,\frac{kl}{x} \, dx$$

or

$$\boxed{\overline{\text{Nu}} = 2\text{Nu}_l = \frac{\bar{h}l}{k}\cdot} \qquad (6\text{-}13)$$

Thus the average value of the Nusselt number is twice the value at the end of the plate $(x = l)$. Of course, we assume here that the flow in the boundary layer is laminar over the entire plate. If turbulence occurs, the above relationships are no longer valid. Turbulence is discussed in a later section.

Now, even for laminar flow, if the Prandtl number is not unity, the foregoing simple expedient does not work. However, approximate solutions can be found rather easily by assuming polynomial temperature profiles in a manner similar to the polynomial velocity profile and using an integral formulation of the thermal boundary layer equation.

6-5 THE THERMAL BOUNDARY–LAYER INTEGRAL EQUATION

Equation (6-4) may be integrated across the boundary layer from 0 to δ_t to yield an integral relationship similar to that of the von Karmann integral relationship for the momentum equation:

$$\int_0^{\delta_t} u \frac{\partial T}{\partial x} dy + \int_0^{\delta_t} v \frac{\partial T}{\partial y} dy = \alpha \int_0^{\delta_t} \frac{\partial^2 T}{\partial y^2} dy$$

It is convenient to replace T by $(T_0 - T)$. Since T_0 will be assumed constant, the equation is the same:

$$\int_0^{\delta_t} u \frac{\partial}{\partial x} (T_0 - T) dy + \int_0^{\delta_t} v \frac{\partial}{\partial y} (T_0 - T) dy = \alpha \int_0^{\delta_t} \frac{\partial^2 (T_0 - T)}{\partial y^2} dy$$

The left-hand side can be rewritten as

$$\int_0^{\delta_t} \frac{\partial}{\partial x} [u(T_0 - T)] dy - \int_0^{\delta_t} (T_0 - T) \frac{\partial u}{\partial x} dy + \int_0^{\delta_t} v \frac{\partial}{\partial y} (T_0 - T) dy$$

and by continuity, $\partial u/\partial x = -\partial v/\partial y$, so that the left-hand side becomes

$$\int_0^{\delta_t} \frac{\partial}{\partial x} [u(T_0 - T)] dy + \int_0^{\delta_t} \frac{\partial}{\partial y} [v(T_0 - T)] dy$$

The second term is

$$\int_{y=0}^{y=\delta_t} d[v(T_0 - T)]$$

and is zero since v is zero at $y = 0$, and $(T_0 - T)$ is zero at $y = \delta_t$. The other term may be written

$$\int_0^{\delta_t} \frac{\partial}{\partial x} [u(T_0 - T)] dy = \frac{d}{dx} \int_0^{\delta_t} u(T_0 - T) dy - \frac{d\delta_T}{dx} u(T_0 - T)\big|_{y=\delta_T}$$

and since $T = T_0$ at $y = \delta_T$, the left-hand side reduces to

$$\frac{d}{dx} \int_0^{\delta_t} (T_0 - T) u\, dy$$

The right-hand term becomes

$$\alpha \int_0^{\delta_t} \frac{\partial^2 (T_0 - T)}{\partial y^2} \, dy = -\alpha \int_0^{\delta_t} \frac{\partial^2 T}{\partial y^2} \, dy = -\alpha \int_{y=0}^{y=\delta_t} d\left(\frac{\partial T}{\partial y}\right) = \alpha \frac{\partial T}{\partial y}\bigg|_{y=0}$$

since $\partial T/\partial y = 0$ at $y = \delta_t$. The final integral form may be written

$$\frac{d}{dx} \int_0^{\delta_t} (T_0 - T) u \, dy = \alpha \frac{\partial T}{\partial y}\bigg|_{y=0} \tag{6-14}$$

which is good for laminar or turbulent flow.

6-6 APPROXIMATE SOLUTION TO THE THERMAL BOUNDARY LAYER

By assuming a polynomial solution for $T(y)$ and making use of Eq. (6-14), a solution may be obtained for $\delta_t(x)$, h_x, and Nu_x for an arbitrary Prandtl number. The solution may be checked with the exact solution of Section 6-4 for Prandtl number unity.

Let us begin by assuming a cubic temperature profile,

$$T = a + by + cy^2 + dy^3$$

The coefficients may be found by the four conditions

$$y = 0 \quad T = T_w \quad \frac{\partial^2 T}{\partial y^2} = 0$$

$$y = \delta_t \quad T = T_0 \quad \frac{\partial T}{\partial y} = 0$$

The condition that $\partial^2 T/\partial y^2 = 0$ at $y = 0$ follows from the energy equation (thermal boundary-layer equation) written at $y = 0$. Evaluating the coefficients, we find

$$\frac{T - T_w}{T_0 - T_w} = \frac{3}{2}\left(\frac{y}{\delta_t}\right) - \frac{1}{2}\left(\frac{y}{\delta_t}\right)^3 \tag{6-15}$$

which is exactly the same form as the cubic velocity profile used in the momentum integral equation. Remember that

$$\frac{u}{U_0} = \frac{3}{2}\left(\frac{y}{\delta}\right) - \frac{1}{2}\left(\frac{y}{\delta}\right)^3 \tag{6-16}$$

By putting the values of T and u from (6-15) and (6-16) into the integral form (6-14), we obtain a differential equation for δ_t in terms of $\delta(x)$ and x. Since we know $\delta(x)$ from the velocity boundary-layer solution, we can ultimately obtain an explicit relationship for $\delta_t(x)$. Equation (6-14) becomes

$$(T_0 - T_w)U_0 \int_0^{\delta_t} \left[1 - \frac{3}{2}\left(\frac{y}{\delta_t}\right) + \frac{1}{2}\left(\frac{y}{\delta_t}\right)^3\right]\left[\frac{3}{2}\left(\frac{y}{\delta}\right) - \frac{1}{2}\left(\frac{y}{\delta}\right)^3\right]dy = \alpha \left.\frac{\partial T}{\partial y}\right|_{y=0}$$

$$= \frac{3\alpha(T_0 - T_w)}{2\delta_t}$$

We shall assume that $\delta_t \leqslant \delta$, so that the integral may be performed in one step. If $\delta_t > \delta$, then u/U_0 will be exactly unity over a portion of the integral between $y = \delta$ and $y = \delta_t$. We shall mention later the result of such a calculation, but for the moment let us assume for simplicity that $\delta_t < \delta$ (Pr $\geqslant 1$). Recall that the solution for $\delta(x)$ using the cubic velocity profile was

$$\delta = \sqrt{\frac{280}{13}}\sqrt{\frac{\nu x}{U_0}} = 4.64\sqrt{\frac{\nu x}{U_0}}$$

although the exact solution was 5.0 instead of 4.64. We shall use 4.64 now, however, to be consistent throughout with our approximate solution. Carrying out the integration, we find

$$U_0 \frac{d}{dx}\left[\frac{3\delta}{20}\left(\frac{\delta_t}{\delta}\right)^2 - \frac{3\delta}{280}\left(\frac{\delta_t}{\delta}\right)^4\right] = \frac{3\alpha}{2\delta_t}$$

We shall neglect the second term of the left-hand side since $(\delta_t/\delta) < 1$ and $3/20 < 3/280$. Then if we let the ratio $\delta_t/\delta = \eta$ and replace δ by $\sqrt{280/13}\sqrt{\nu x/U_0}$, we have

$$\eta^3 + \frac{4}{3} \times \frac{d}{dx}(\eta^3) = \frac{13}{14Pr} \tag{6-17}$$

It can readily be verified that the general solution to this equation is a particular integral $13/14Pr$ and a solution to the homogeneous equation $Ax^{-3/4}$. Hence

$$\eta^3 = \left(\frac{\delta_t}{\delta}\right)^3 = \frac{13}{14Pr} + Ax^{-3/4} \tag{6-18}$$

The boundary condition to determine the constant A is a bit tricky. Perhaps the easiest way to determine A is as follows. Remember that for Pr $= 1$ we found previously that $\delta_t = \delta$ and $\eta = 1$. Since Eq. (6-18) must reduce to no dependence on x when $\Pr = 1$, we see that A must be zero and

$$\eta = \frac{\delta_t}{\delta} = \frac{1}{1.026\sqrt[3]{\Pr}}$$

(6-19)

The fact that δ_t/δ is not exactly unity for $\Pr = 1$ is due to the fact that we have an approximate solution based on the cubic temperature and velocity profiles.

Now, h_x and Nu_x can be determined from our value of δ_t and the profile (6-15):

$$q = -k \left.\frac{\partial T}{\partial y}\right|_{y=0} = h(T_w - T_0) = k \frac{3(T_w - T_0)}{2\delta_t}$$

$$h_x = \frac{3k}{2\delta_t} = \frac{3k}{2\delta\eta}$$

and using δ as $4.64\sqrt{\nu x/U_0}$, we obtain

$$\boxed{h_x = 0.332k\sqrt[3]{\Pr}\sqrt{\frac{U_0}{\nu x}}}$$

(6-20)

and the corresponding Nusselt number Nu_x is

$$\boxed{\mathrm{Nu}_x = \frac{(h_x)(x)}{k} = 0.332\sqrt[3]{\Pr}\sqrt{\mathrm{Re}_x}}$$

(6-21)

It is interesting to note that this expression is the same form as the solution for $\Pr = 1$ except now we see that the Prandtl number dependence is simply $\sqrt[3]{\Pr}$. Of course, this result is good only for $\Pr > 1$ or near unity. Gases, as we said, have Prandtl numbers near unity, and most liquids have $\Pr > 1$.

However, liquid metals have rather small Prandtl numbers ($\Pr \ll 1$), and Eq. (6-19) is not valid for $\Pr$ much less than one. The integration of Eq. (6-14) must then be carried out in two steps, one for $0 < y < \delta$, where u/U_0 is given by the cubic profile, and one for $\delta < y < \delta_t$, where u/U_0 is unity. We shall not carry out the details here, but the final result for Nu_x is

$$\mathrm{Nu}_x = \frac{\sqrt{\mathrm{Re}\,\Pr}}{1.55\sqrt{\Pr} + 3.09\sqrt{0.372 - 0.15\Pr}}$$

(6-22)

The appropriate values for $\bar{h}$ and $\overline{\mathrm{Nu}}$ for a plate of length l can be found, and the relationships between h_x and $\bar{h}$ and Nu_x and $\overline{\mathrm{Nu}}$ are exactly the same as given by Eqs. (6-12) and (6-13).

6-7 THE TURBULENT BOUNDARY LAYER AND THE REYNOLDS ANALOGY

The film coefficient and Nusselt number have the same significance in turbulent flow as in laminar flow, and in fact the expressions in turbulent flow are probably more important from a practical point of view since most flows of engineering interest are turbulent.

The same problems arise here as in the determination of the velocity profile in turbulent flow. A temperature sublayer exists, and some recourse must be made to experiments.

In laminar flow, heat is conducted from (or to) the wall into the fluid and convected along the flow direction by the moving fluid in the form of heat stored in the heated fluid. The conduction process dominates near the wall, and out into the boundary layer the velocity increases and the convection becomes more important. In turbulent flow, however, the turbulent fluctuations in fluid velocity cause mixing of the heated fluid and a consequent transfer of heat by this "turbulent mixing" process. Analytical calculations cannot be carried out from first principles in turbulent flow. The details of the processes involved are still not completely understood. Some approximations are necessary in order to achieve a usable solution to the problem.

We can start with Eq. (6-14), the integral formulation, and insert the appropriate velocity profiles for turbulent flow. But then the appropriate temperature profile is unknown and cannot be represented by the simple polynomial as in laminar flow.

An important method for finding a relationship between τ_w (the wall shear) and q_w, and consequently h and Nu for turbulent flow when $\text{Pr} = 1$, was shown by Reynolds (1874) and later extended by Prandtl, G. I. Taylor, and others.

Remember that in the laminar region (and in the laminar sublayer even in turbulent flow) the shear τ and heat flux q are related by a simple expression if $\text{Pr} = 1$. Referring to Section 6-4, we saw that for $\text{Pr} = 1$ in laminar flow, q_w and τ_w were related by

$$q_w = \frac{k}{\mu U_0}(T_w - T_0)\tau_w = \frac{c_p}{U_0}(T_w - T_0)\tau_w$$

But throughout the laminar flow a more general differential relationship can be established that holds not just at the wall and is good for arbitrary Prandtl number. From the definitions,

$$\tau = \mu \frac{\partial u}{\partial y}$$

$$q = -k \frac{\partial T}{\partial y}$$

we can say that (for a fixed value of x) τ and q in laminar boundary layer flow are related by

$$q = -\tau \frac{k}{\mu} \frac{dT}{du} \qquad (6\text{-}23)$$

We can establish a similar relationship for turbulent flow (outside the laminar sublayer). In turbulent flow we can neglect the thermal conduction in the fluid compared to the turbulent mixing and consequent transport of heat due to turbulent fluctuation. Similarly, the viscous shear stress may be neglected compared to the apparent turbulent stresses (or Reynolds stresses as they are called).

Consider a parallel plane A–A, parallel to the wall, in the turbulent flow region (Fig. 6-4). By the turbulent mixing process the fluid moves upward through the plane A–A, and in steady flow the same amount of fluid moves downward. Over a given area on the plane A–A, little bits of fluid continuously move up and down through the plane in a random fashion. If we follow a group of tiny blobs of fluid of total mass m per unit area (of A–A) per unit time as they move from some level at y denoted as 1–1 below A–A to a level $y + \Delta y$ denoted as 2–2 above A–A, it transports a bit of heat (actually, we consider transport of enthalpy if it is a gas and internal energy if a liquid). The fluid blobs bring up to A–A the enthalpy

$$mc_p T_1$$

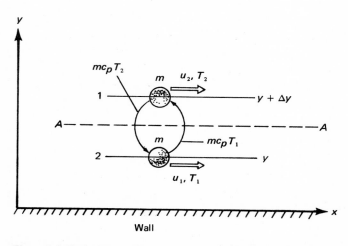

Figure 6-4 Turbulent exchange in the turbulent boundary layer.

and an equal mass moving down transports

$$mc_pT_2$$

The net amount transported *up* away from the wall (positive in the sense of positive q) is

$$q = mc_p(T_1 - T_2) \tag{6-24}$$

per unit area per unit time.

Similarly, the shear stress (on the plane A–A) due to the turbulence may be found and the viscous shear $\mu(\partial u/\partial y)$ neglected. This "turbulent shear stress" is actually an apparent stress due to the momentum exchange associated with the turbulent exchange of fluid across the plane. Consider a control volume of unit cross-sectional area (parallel to the wall) as shown in Fig. 6-5. The net momentum flux up through the area A–A is then

$$m(u_1 - u_2)$$

and this is equivalent to an apparent turbulent shear stress τ. Hence

$$\tau = -m(u_1 - u_2) \tag{6-25}$$

The minus sign is necessary because τ is positive in the positive x direction and is an equivalent stress, not one that balances $m(u_1 - u_2)$. Taking the ratio of (6-24) to (6-25), we have

$$q = -\tau c_p \frac{T_1 - T_2}{u_1 - u_2} = -\tau c_p \frac{T_y - T_{y+\Delta y}}{u_y - u_{y+\Delta y}}$$

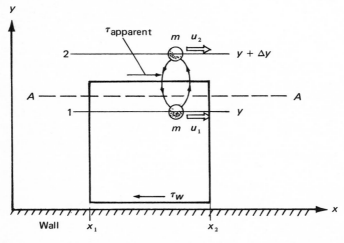

Figure 6-5 The apparent shear stress due to turbulent exchange across plane A–A.

and in the limit as $\Delta y \to 0$ we have the equation known as Reynolds' analogy,

$$q = -\tau c_p \frac{dT}{du} \tag{6-26}$$

Now, Eqs. (6-23) and (6-26) are identical and hold both in laminar and turbulent flow only if $k/\mu = c_p$, that is, $Pr = 1$. Hence when $Pr = 1$ the shear stress and heat flux rate are related by the same equation in laminar and turbulent flow.

Now, we know the shear stress τ and profile $u(y)$ in both laminar and turbulent flow from our study in Chapter 5. Thus we can extend our equation for h and Nu for laminar flow (and $Pr = 1$) to the turbulent boundary layer for $Pr = 1$.

We can integrate Eq. (6-26) from the wall out to any arbitrary point in the turbulent boundary layer. (We don't do this for laminar flow because we already have those results.) In particular, it is convenient to integrate from $y = 0$ (the wall) to $y = \delta$ (the free-stream value):

$$\int_{T_w}^{T_0} dT = -\frac{1}{c_p} \int_0^{U_0} \frac{q}{\tau} \, du$$

Since in laminar flow q/τ is not a function of y and is just equal to q_w/τ_w, we make the assumption that q/τ remains approximately uniform with y up through the turbulent boundary layer. Then we have

$$T_0 - T_w = -\frac{q_w}{c_p \tau_w} U_0$$

or

$$q_w = \frac{c_p \tau_w}{U_0} (T_w - T_0) = h_x (T_w - T_0) \tag{6-27}$$

which is identical to the expression for laminar flow. Now the appropriate expression for τ_w from Chapter 5 may be used for turbulent flow. Using expression (5-25), $\tau_w = 0.0296 \rho U_0^2 (Re_x)^{-1/5}$, we readily obtain for h_x and Nu_x

$$\frac{h_x}{\rho c_p U_0} = (0.0296)(Re_x)^{-1/5}$$

$$\tag{6-28}$$

$$Nu_x = \frac{(h_x)(x)}{k} = 0.0296(Re_x)^{0.8}$$

The mean values for a plate of length l (assuming the boundary layer to be turbulent from the leading edge, which may be a fairly good assumption for long plates) are

$$\bar{h} = \frac{1}{l} \int_0^l h_x \, dx = \frac{(0.0296)\rho c_p U_0}{l} \int_0^l \left(\frac{U_0 x}{\nu}\right)^{-1/5} dx$$

so that

$$\left.\begin{array}{c} \bar{h} = \dfrac{5}{4}h_l \\[2mm] \overline{\mathrm{Nu}} = \dfrac{\bar{h}l}{k} = \dfrac{\frac{5}{4}h_l l}{k} = \dfrac{5}{4}\mathrm{Nu}_l \end{array}\right\} \qquad (6\text{-}29)$$

Remember, however, that these expressions are good only for Pr $= 1$.

For a general value of Pr, the laminar sublayer and turbulent region must be treated separately. We shall not go through the analysis here, but the results are not too difficult to obtain. They are

$$\frac{h_x}{\rho c_p U_0} = (0.0296)(\mathrm{Re}_x)^{-1/5}(\mathrm{Pr})^{-2/3}$$

$$\mathrm{Nu}_x = (0.0296)(\mathrm{Re}_x)^{0.8}(\mathrm{Pr})^{1/3} \qquad (6\text{-}30)$$

And as for Pr $= 1$, the relationship between h_x and $\bar{h}$ and Nu_x and $\overline{\mathrm{Nu}}$ are the same: $\bar{h} = \frac{5}{4}h_l$, $\overline{\mathrm{Nu}} = \frac{5}{4}\mathrm{Nu}_l$.

It is interesting to note that the expression $h_x/\rho c_p U_0$, which appears in Eqs. (6-28) and (6-30), is a dimensionless parameter that occurs in turbulent heat transfer. It is known as the *Stanton number* and is written as St. We see that St is simply the ratio Nu/Re Pr. Equation (6-30) is usually written in the form

$$\boxed{\mathrm{St} = 0.0296(\mathrm{Re}_x)^{-1/5}(\mathrm{Pr})^{-2/3}} \qquad (6\text{-}31)$$

We have briefly studied the flat plate in longitudinal flow. In engineering practice we often are interested in flow in more complex configurations, such as flow in and over tubes, both axially and across; flow with pressure gradients and separation; and the study of simultaneous flow streams on different sides of plates in heat exchangers. Theory gets difficult in such situations, but experiment can establish empirical relationships between the relevant parameters. Such empirical "correlations" are of utmost importance in heat transfer. They usually take the form of equations or curves of

$$\mathrm{Nu} = f(\mathrm{Re}, \mathrm{Pr})$$

or

$$St = f(Re, Pr)$$

Now, remember that we have been concerned so far with "forced" convection; that is, the fluid was forced over the wall with velocity U_0. We can now turn our attention to "natural" or "free" convection, an important mechanism in both heating or cooling in situations where no fan or pump is used.

6-8 FREE CONVECTION

"Natural" or "free" convection takes place when the fluid motion is caused by buoyancy effects. If the fluid near the wall becomes heated or cooled sufficiently compared to the bulk of the fluid, the consequent density differential generates a buoyant force near the wall and a velocity and thermal boundary layer. The velocity and thermal layers are intimately tied together, since the thermal effect must be present to drive the fluid. The velocity and temperature profiles are coupled, and the momentum and energy equations must be solved simultaneously.

Of course, even in forced convection the temperature near the wall will be different from the free-stream or bulk temperature of the fluid. However, buoyancy effects are usually negligible in forced convection, and any density difference throughout the fluid is neglected, as in the previous sections in which we discussed forced convection. In forced convection, fluid is forced over an object as, for example, by blowing, by moving the object through the fluid, or by causing motion by an imposed pressure gradient as in a pipe or duct. But in free convection such imposed or forced motions are absent or negligible. Hence, in order for free convection to dominate, we must have a situation where motion would not take place without the heating effect and the physical configuration must be such as to allow the fluid movement.

The simplest configuration is a heated vertical wall, which generates an upward movement of fluid near the wall as it is heated. The common household hot water "radiator" (a misnomer actually) is an example of a vertical wall convecting to room air. Cooling fins on hot objects (such as transistor mounts, air-cooled engine cylinders, etc.) are essentially hot vertical plates with free convection boundary layers. A common example of hot air rising is the air in a chimney. The hot air rises up the chimney because it is less dense than the surrounding air and buoyancy forces move the air up the chimney.

Although we shall discuss the vertical heated plate here in some detail, there are many other examples of free convection that are more complicated and that we can mention only in passing. Important examples are the free convection from a heated horizontal flat plate and objects other than flat plates. Much theoretical and experimental work has been done on these problems, which are rather important from an engineering standpoint.

A horizontal flat plate will not generate a boundary layer. A heated plate causes the fluid on top to become warmer than the surrounding fluid. At first nothing happens, but eventually the density of the warmer fluid becomes small enough to generate a large enough buoyancy force to cause the warm fluid to "tunnel" upward and overturn the fluid. Hexagonal convection cells, called *tessellation* patterns, are formed. These cells are vertical cylinders of roughly hexagonal cross-sectional shape which extend upward from the heated plate. The fluid rises upward and returns back down in a toroidal or ring-type motion within each cell. These cells are known as *Bènard cells.* Such a problem is much more complex than the boundary layer on a vertical plate and involves the stability of the horizontal layer or statification of the density above the plate.

Furthermore, there are instances where both free and forced convection effects are important, as when, for example, the natural convection flow on a vertical plate is augmented by blowing. Again the analysis is rather complicated, and we shall not pursue such problems.

Let us now examine the vertical plate.

The Vertical Flat Plate

We shall treat the vertical plate by the integral approximation method just as we did forced convection flow. Now, however, we must solve the momentum and energy equations together. We can solve for both laminar and turbulent flow, but we shall carry out the details only for laminar flow. Referring to Fig. 6-6, we see a sketch of the temperature and velocity profiles. The velocity is zero both at the wall ($y = 0$) and in the free stream ($y > \delta$) and rises to a maximum somewhere in between. The boundary-layer thickness δ may be assumed to be the same for temperature and velocity. This assumption is not quite correct, but a more detailed analysis shows that the difference is negligible for most engineering work.

The integral form of the momentum equation must now take into account the buoyancy body force. The boundary-layer equation of motion taking into account the gravitational body force (per unit volume) $-\rho(\partial\psi/\partial x)$ in the x direction may be written, where ψ is the gravitational potential* gx, and $-\rho g = -\rho(\partial\psi/\partial x)$.

---

*To be consistent with conventional notation, we have used ψ as the symbol both for gravitational potential and stream function, but no confusion should arise since it is explained each time it appears.

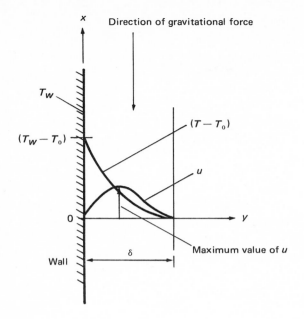

Figure 6-6 The boundary layer due to free convection along a heated vertical plate. The values of δ and δ_t are approximately the same. The gravitational force is assumed to act in the negative x direction.

$$\rho\left(u\frac{\partial u}{\partial x} + v\frac{\partial u}{\partial y}\right) = -\frac{\partial P}{\partial x} + \mu\frac{\partial^2 u}{\partial y^2} - \rho\frac{\partial \psi}{\partial x} \qquad (6\text{-}32)$$

The only pressure gradient is due to the hydrostatic pressure differential due to elevation. But the density ρ variation with temperature may be expressed as

$$\beta = -\frac{1}{\rho}\left(\frac{\partial \rho}{\partial T}\right)_P$$

or for small temperature variations,

$$\Delta\rho = -\rho\beta\,\Delta T \qquad (6\text{-}33)$$

where β is the temperature coefficient of expansion. For a gas the β is replaced by $1/T$ and

$$\Delta\rho = -\rho\frac{\Delta T}{T} \qquad (6\text{-}34)$$

If we assume that changes in ρ and T are small in the boundary layer, we can write the values of ρ and T as the sum of a small perturbation denoted with a prime and a large background value denoted as a nought. Hence,

$$\rho = \rho' + \rho_0 \qquad \rho' \ll \rho_0$$

$$T = T' + T_0 \qquad T' \ll T_0$$

The pressure is assumed not to vary with y and hence is simply $P_0(x)$. This is consistent with the boundary-layer assumptions. Substituting these values into (6-32) and (6-34), we have

$$(\rho' + \rho_0)\left(u\frac{\partial u}{\partial x} + v\frac{\partial u}{\partial y}\right) = -\frac{\partial}{\partial x}(P_0) - (\rho' + \rho_0)g + \mu\frac{\partial^2 u}{\partial y^2}$$

$$\rho' = -(\rho' + \rho_0)\beta(T')$$

This equation may be separated into a zero-order balance of nought (or free-stream) quantities that holds in the free stream and a first-order equation for the boundary layer:

$$0 = -\frac{\partial P_0}{\partial x} - \rho_0 g$$

which gives the hydrostatic pressure variations, and

$$\rho_0\left(u\frac{\partial u}{\partial x} + v\frac{\partial u}{\partial y}\right) = -\rho'g + \mu\frac{\partial^2 u}{\partial y^2}$$

for the first-order balance. The equation of state is

$$\rho' = -\rho_0\beta T' = -\rho_0\beta(T - T_0)$$

for a liquid, or

$$\rho' = -\frac{\rho_0}{T_0}T'$$

for a gas. Combining, we have

$$\boxed{\left(u\frac{\partial u}{\partial x} + v\frac{\partial u}{\partial y}\right) = \beta(T - T_0)g + v\frac{\partial^2 u}{\partial y^2}}\tag{6-35}$$

where $v = \mu/\rho_0$. The energy equation remains unchanged, simply Eq. (6-4).

The momentum integral equation may be found by integrating (6-35) over the boundary layer just as we did before in Chapter 5 [Eq. (5-5)]. The only difference now is an additional term $g\beta \int_0^\delta (T - T_0)\,dy$ and the fact that $U_0 = 0$. The result is

$$\frac{d}{dx}\int_0^\delta u^2\,dy = g\beta\int_0^\delta (T - T_0)\,dy - v\frac{du}{dy}\bigg|_{y=0}\tag{6-36}$$

and the energy equation is unchanged and remains

$$\frac{d}{dx}\int_0^\delta u(T - T_0)\,dy = -\alpha \left.\frac{dT}{dy}\right|_{y=0} \tag{6-37}$$

Polynomial approximations may be assumed for the temperature and velocity profiles and the integral equations solved together for $\delta(x)$.

Even the approximate solution is a bit involved, however, and a numerical solution is almost as practical. By appropriate similarity transformations the equations (6-4) and (6-35) may be reduced to two simultaneous ordinary non-linear equations, which may be solved numerically. If we let

$$\eta = \frac{y}{x}\left(\frac{\mathrm{Gr}_x}{4}\right)^{1/4}$$

where Gr_x is a dimensionless number that occurs in natural convection, the Grashof number, defined as $\mathrm{Gr}_x = g\beta(T_w - T_0)x^3/\nu^2$ for liquids and as $\mathrm{Gr}_x = g(T_w - T_0)x^3/T_0\nu^2$ for gases. We introduce a stream function ψ (as we did for the Blasius solution in Chapter 5) as $u = \partial\psi/\partial y$, $v = -\partial\psi/\partial x$:

$$\psi = 4\nu\zeta(\eta)\left(\frac{\mathrm{Gr}_x}{4}\right)^{1/4} \qquad u = \frac{4\nu\zeta'}{x}\left(\frac{\mathrm{Gr}_x}{4}\right)^{1/2} \qquad v = \frac{\nu}{x}\left(\frac{\mathrm{Gr}_x}{4}\right)^{1/4}(\eta\zeta' - 3\zeta)$$

where ζ' indicates $d\zeta/d\eta$. Then the governing equations become

$$\frac{d^3\zeta}{d\eta^3} + 3\zeta\frac{d^2\zeta}{d\eta^2} - 2\left(\frac{d\zeta}{d\eta}\right)^2 + \theta = 0$$

$$\frac{d^2\theta}{d\eta^2} + 3\mathrm{Pr}\zeta\frac{d\theta}{d\eta} = 0 \tag{6-38}$$

where $\theta = (T - T_0)/(T_w - T_0)$. Numerical solutions were first carried out by Pohlhausen (1930).[*] For air ($\mathrm{Pr} = 0.714$), a value of Nu_x was obtained as

$$\boxed{\mathrm{Nu}_x = \frac{hx}{k} = 0.360(\mathrm{Gr}_x)^{1/4}} \tag{6-39}$$

Numerical calculations using a computer were carried out by Ostrach (1953)[†] and are shown in Fig. 6-7.

We shall not tabulate the results of Nu_x for arbitrary Prandtl number, but

---

[*]In collaboration with E. Schmidt and W. Beckman, *Forsch-Ing.-Wes.*, vol. 1, p. 391, 1930

[†]S. Ostrach, *ASME Trans.*, vol. 75, p. 1287, 1953.

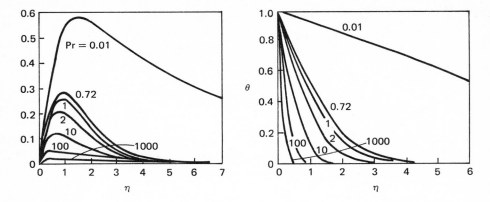

Figure 6-7 Temperature and velocity profiles in laminar natural convection for various Prandtl numbers. (After Ostrach, 1953.)

for simple calculations the result of a polynomial approximation (quadratic in temperature) do give an explicit expression as

$$\text{Nu}_x = 0.508 \text{Pr}^{1/2}(0.952 + \text{Pr})^{-1/4}(\text{Gr}_x)^{1/4} \tag{6-40}$$

which reduces to

$$\text{Nu}_x = 0.378(\text{Gr}_x)^{1/4}$$

for air. The number 0.378 checks quite well with the more exact value 0.360. The average value of h over a plate of height l can be found by integrating (6-39) or (6-40) to get

$$\bar{h} = \tfrac{4}{3}h_l \tag{6-41}$$

As a matter of engineering "feel," h values of about 1 to 10 Btu/h $\cdot$ ft^2 $\cdot$ °F in air might be expected in ordinary natural convection situations as compared to h values of one to several orders of magnitude greater for forced convection.

6-9 EFFECTS OF FRICTIONAL HEATING IN THE BOUNDARY LAYER

So far in this chapter we have assumed that friction in the boundary layer is unimportant and is negligible insofar as it affects the heat flow between the wall and the free stream. However, at very large free-stream velocities, the friction may become important. If significant heat is generated in the boundary layer by friction, the fluid near the wall may become hotter than either the wall or the free stream. In such cases the direction of heat transfer near the wall can be changed if friction becomes important even though the temperature difference $(T_w - T_0)$ remains the same. In the cooling of a plate, for example, the plate may cease to be cooled and actually heated by a high-speed free stream. Such is the case in high-speed aerodynamic flow, particularly in supersonic flight, where a cold-air free stream actually causes heating of the airplane surface because of the friction in the boundary layer.

A flat plate inserted in a free stream at velocity U_0 and temperature T_0 will rise to the adiabatic wall temperature T_a (analogous to the adiabatic wall temperature discussed in Chapter 4 for Couette flow). If we assume the plate to be at a constant temperature T_a, the value of T_a may be found from a solution to the thermal boundary layer including frictional effects. Such a solution can be found numerically from the equation for the incompressible thermal boundary layer. Equation (6-6) with the frictional heating term [from Eq. (6-3)] left in can be expressed as

$$\frac{d^2T}{d\eta^2} + \frac{\mathrm{Pr}}{2}f\frac{dT}{d\eta} = -\mathrm{Pr}\frac{U_0^2}{c_p}(f'')^2 \tag{6-42}$$

where f is the same as in Eq. (6-6). The solution is a function not only of Prandtl number but also of another parameter that depends on the viscous dissipation and involves U_0^2/c_p, the Eckert number, defined as $\mathrm{E} = U_0^2/c_p(T_w - T_0)$. Here E is exactly the same as the E used in Chapter 4 where we discussed frictional effects in Couette flow. Again we shall not go through the solution, but the results for air (Pr = 0.7) are shown in Fig. 6-8. We see that for $\sqrt{\mathrm{Pr}}\,\mathrm{E} > 2$, the wall is actually heated even though $T_0 < T_w$.

We see that for $q_w = 0$, $dT/d\eta = 0|_{\eta=0}$ and $\sqrt{\mathrm{Pr}}\,\mathrm{E} \approx 2$, so that

$$\frac{\sqrt{\mathrm{Pr}}\,U_0^2}{c_p(T_a - T_0)} \approx 2$$

This gives the adiabatic wall temperature T_a as the temperature at which the wall "floats" in the stream as

$$T_a - T_0 \approx \frac{\sqrt{\mathrm{Pr}}\,U_0^2}{2c_p} \tag{6-43}$$

This result is exact only for Pr close to unity (good for air). For Pr far from unity the parameter $\sqrt{\mathrm{Pr}}\,\mathrm{E}$ in Fig. 6-8 must be modified to $F(\mathrm{Pr})\,\mathrm{E}$, where F is a function that is tabulated along the way when the solution to Eq. (6-42) is found. For aerodynamic work, Eq. (6-43) is generally used.

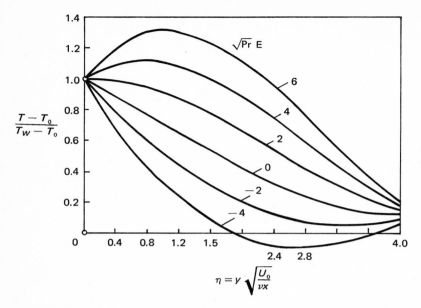

Figure 6-8 Temperature profile in a boundary layer on a flat plate including frictional heating effects (for air, Pr ≈ 0.7). E > 0 indicates heating and E < 0 indicates cooling of the plate (after Schlichting, 1964).

Another way to think of the adiabatic wall temperature is that heat will flow from a hot plate to a cooling stream $(T_w > T_0)$ only if $T_w > T_a$. Or, if $T_w < T_0$, heat flows to the wall only if $T_w < T_a$.

Although Eq. (6-43) was derived on the basis of incompressible flow, the effect of compressibility is rather slight and Eq. (6-43) may be used to a high degree of approximation in aerodynamic work. In compressible flow, Eq. (6-43) is now often written in a somewhat more general form:

$$T_a - T_0 = r \frac{U_0^2}{2c_p} \tag{6-44}$$

where r is known as the *recovery factor* and physically represents the ratio of the frictional temperature increase $(T_a - T_0)$ to the temperature increase due to adiabatic compression, $U_0^2/2c_p$. This term $U_0^2/2c_p$ is the difference between the stagnation temperature T_s and the free-stream temperature T_0. The energy equation along a streamline is $U_0^2/2 + c_p T_0 = \text{constant} = c_p T_s$ for adiabatic flow (even with friction), so that $T_s - T_0 = U_0^2/2c_p$. In air $r \approx \sqrt{\text{Pr}} \approx 0.85$.

If we introduce the free-stream Mach number, $M_0 = U_0/C_0$ (where C_0 is the velocity of sound in the free stream), which may be written for a perfect gas as $C_0^2 = \gamma R T_0 = (\gamma - 1)c_p T_0$ (where R, the gas constant, is $c_p - c_v$), we have

$$T_a = T_0\left(1 + \sqrt{\text{Pr}}\,\frac{\gamma - 1}{2}M_0^2\right) \tag{6-45}$$

In other words, in a compressible fluid, the stagnation temperature difference $U_0^2/2c_p$ must be multiplied by r to give the adiabatic wall temperature.

6-10 AERODYNAMIC HEATING

We close this chapter with a brief discussion of a problem that assumed great importance during the flourishing of the manned space program. Vehicles traveling at very high speed through the atmosphere become heated because of viscous friction and the surface temperature, or, in effect, the adiabatic wall temperature may rise to a high level—beyond the endurance of man.

At supersonic speeds, say Mach number 1 to 2, the adiabatic wall temperature gives us an idea of the temperature of the airplane surface in steady flight. At $M = 2$, for example, $T_a/T_0 \approx 1.8$ (for an air temperature of $0°F$, T_a is about $370°F$). These high temperatures do constitute a problem, and in commercial supersonic jet planes the maintaining of the inside temperature at a reasonable level is aggravated by the high outside surface temperature, even though the outside air may be quite cold. Proper insulation and air conditioning are vital.

For example, a reentry vehicle may enter the upper atmosphere at hypersonic speed with a Mach number of about 10 to 30. At these speeds the temperature in the boundary layer is so high that dissociation of the air molecules and even ionization occurs. The complete description of the boundary layer must take these chemical and atomic reactions into account. The behavior of hypersonic boundary layers and wakes is rather complex, and it is only recently, in connection with space vehicle reentry, that a detailed analytical and experimental study has been made.

However, we can apply the results of our thermal boundary layer studies in this chapter to the hypersonic heating problem and at least obtain some interesting qualitative information about how a reentry vehicle should be designed so that it does not burn up while passing down through the atmosphere. The conclusions are quite different from those for an airplane. For an airplane, the drag must be minimized and reasonable steady temperatures maintained. For a reentry vehicle, the problem is basically a transient one, and drag need be adjusted to minimize the heating—not to minimize power since a reentry vehicle is basically in free fall.

For an order-of-magnitude study we can express Newton's law of motion for the vehicle as

$$M\frac{dV}{dt} = -D = -C_D(\tfrac{1}{2}\rho V^2 A)$$

or

$$M\frac{d(V^2/2)}{dt} = -DV$$

where M is the mass of the vehicle, V is its velocity, D is the drag opposing the motion, and A is a characteristic surface area. (We assume gravity to be small compared to the drag.)

Now, the total rate of heat transfer dQ/dt to the vehicle at temperature T_w may be expressed approximately in terms of the adiabatic wall temperature and T_w as

$$\frac{dQ}{dt} \approx \bar{h}A(T_A - T_w) \approx \frac{\bar{h}AV_0^2}{2c_p}$$

For air, $T_A \approx T_0 + (rV_0^2/2c_p)$, where T_0 and V_0 are the free-stream values. We assume that $r \approx 1$ and conservatively we assume that $T_w \approx T_0$. Further, from Eq. (6-9), it follows that

$$\bar{h} \propto \tfrac{1}{2}\rho c_p C_{Df} V_0$$

where the drag due to friction D_f is

$$D_f = C_{Df}\frac{\rho V_0^2}{2}A$$

which defines C_{Df} as the skin friction coefficient. The total drag D may be written

$$D = C_D \frac{\rho V_0^2}{2}$$

$$C_D = C_{Df} + C_{DP}$$

where C_{D_p} is the pressure or form drag coefficient. Combining these relationships, we have

$$\frac{dQ}{dt} \approx -\frac{1}{2}\frac{C_{Df}}{C_D}M\frac{d(V_0^2/2)}{dt}$$

and hence Q, the total heat transferred to the object during deceleration, is the integral of the above equation and may be expressed in terms of the initial velocity V_{0i} as

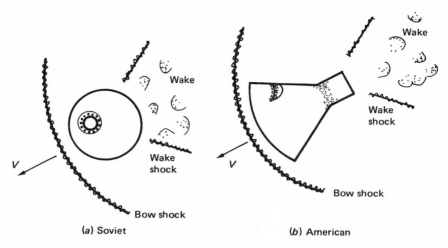

Figure 6-9 Typical reentry vehicle shapes.

$$Q \approx \frac{1}{2} \frac{C_{Df}}{C_D} \left(\frac{MV_{0i}^2}{2} \right) \tag{6-46}$$

where the final velocity is neglected. We see that in order to minimize the heat input to the object or spacecraft, the ratio (C_{Df}/C_D) should be small. In other words, the object should be blunt, with the form drag making up the main contribution to the total drag. Even then, ablative materials are used on the surface of reentry vehicles to help absorb the heat and prevent the interior temperature from rising too high. In the Soviet Union, typical reentry vehicles have been spherical, whereas American craft have generally been blunt but conical in shape (Fig. 6-9). The sphere has a lower value of C_{Df}/C_D, but the cone shape has better directional stability during the reentry flight.

PROBLEMS

6-1 The laminar boundary layer over a flat plate may be stabilized at constant thickness by sucking the fluid through the plate, which must be porous. Boundary-layer suction is often used to prevent separation.

Consider such a flow with a uniform suction velocity V_0 down through the plate (Fig. 6-10). In the fully developed flow, find the expressions for the velocity profile, the film coefficient, and the Nusselt number, assuming the Prandtl number to be arbitrary. The plate is held at temperature T_w and the free stream is T_0. Is $\delta = \delta_t$?

6-2 The arrangement sketched in Fig. 6-11 (a hot plate anemometer) is to be used to measure the velocity of a stream of air, either the mean velocity or instantaneous velocity

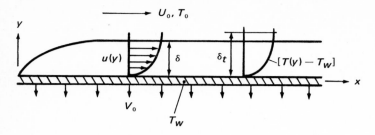

Figure 6-10

with limits on resolution set by the plate size. A wide, thin, flat plate of length l and projected area A is inserted in the air stream parallel to the flow. The air flows turbulently past both sides of the plate with a velocity U_0 and a known temperature T_0. The velocity is low enough so that frictional heating is negligible. The leading and trailing edges of the plate are electrodes through which a current i, supplied by a battery at a voltage E, passes along the plate. The rheostat is adjusted so that the current remains nearly constant with time. The rheostat resistance is made much larger than that of the plate. The resistance of the plate varies with temperature according to

$$R = R_0 [1 + \alpha(T - T_0)]$$

where R_0 and α are known.

Assuming that the temperature is uniform thoughout the plate, relate the unknown velocity U_0 to the measured current and voltage and other constants of the system.

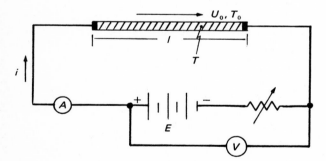

Figure 6-11

6-3 Derive Eq. 6-34, $\Delta\rho = -\rho(\Delta T/T)$. What assumptions are necessary for this equation to be a valid approximation?

Answer: The condition that $\Delta P/P_0 \ll \Delta T/T_0$ must hold.

Show that this condition can indeed be satisfied in a vertical heated column or boundary layer of gas.

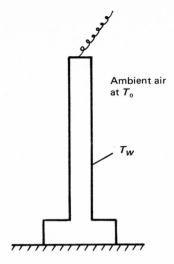

Ambient air
at T_0

T_W

Figure 6-12

6-4 Consider the flow upward through a tall circular chimney, as shown in Fig. 6-12. In order to estimate the flow rate of gas up the chimney, we may make the following assumptions: (1) The wall of the chimney is at a uniform temperature T_W and the outside air is at a uniform temperature T_0. (2) The flow in the chimney is fully developed, both in velocity and temperature. (3) The viscous dissipation is negligible.

Determine the velocity profile assuming the flow to be laminar.

Hint: If the temperature profile is fully developed, the temperature must be uniform throughout the fluid. Is this temperature T_W? Can the density then be assumed to be constant throughout the fully developed flow region?

6-5 Calculate the Nusselt numbers and average Nusselt numbers for a flat plate of length L placed in a stream of water (parallel to the velocity of the water) for various values of L (1 m, 5 m, 10 m) in various combinations of various values of U_0 (1 m/s, 10 m/s, 50 m/s). Assume that the critical value of Re_x for transition from laminar to turbulent is 10^5, and for simplicity that the boundary layer is either all laminar or all turbulent.

Repeat the calculations for air at standard conditions.

6-6 For Problem 6-5, find the rates of heat transfer from the plate to the free stream if the plate is held at a temperature of 80°C and the free stream is at 20°C.

6-7 The plate of Problem 6-5 is now rotated so that it is vertical and placed in a natural environment of still air. The plate is heated to 100°C (the air is at 20°C and standard pressure). Calculate the rate of heat transfer from the plate to the air for a plate height of 1 m, 10 m, and 100 m. How do these values compare to those of the previous problem?

Find the mean film coefficient $\bar{h}$ (in English engineering units) for each plate height. A rule of thumb is that $\bar{h}$ for a vertical flat plate in air is about unity (in English units). Is this a good rule? Discuss.

BIBLIOGRAPHY

General

Eckert, E. R. G., and R. M. Drake, Jr.: "Analysis of Heat and Mass Transfer," McGraw-Hill Book Company, New York, 1972.

Rohsenow, W. M., and H. Choi: "Heat, Mass, and Momentum Transfer," Prentice-Hall, Englewood Cliffs, N.J., 1961.

Schlicting, H.: "Boundary Layer Theory," 4th ed., McGraw-Hill Book Company, New York, 1961.

Historical

Pohlhausen, E.: *Z. Math. Mech.*, vol. 1, p. 115, 1921.

SOME BASIC EQUATIONS IN
VARIOUS COORDINATE SYSTEMS

STRAIN RATE RELATIONSHIPS

The deformation rate tensor is written (in Cartesian tensor notation) as $\partial w_i/\partial x_j$, where w_i is the velocity. The symmetrical part of the deformation rate tensor is the strain rate tensor, and the antisymmetrical part is the rotation tensor.

The strain rate tensor is denoted as e_{ij} and the rotation tensor as ω_{ij}. The deformation rate tensor can then be written in terms of e_{ij} and ω_{ij} as $\partial w_i/\partial x_j = e_{ij} + \omega_{ij}$ in Cartesian tensor notation. The components of the strain rate tensor and rotation tensor are given in various coordinate systems, so that the deformation rate tensor can be found by adding them together.

The rotation term can be related to the angular velocity ω_j of an infinitesimal fluid element as $\omega_j = \omega_{ik}$ and is indicated below.

The diagonal (normal) components of the strain rate tensor can be identified directly with the true normal strain rate. However, the off-diagonal terms (shear rate components), e_{ij}, $i \neq j$, are equal to one-half the true rate of shear strain components, which are denoted as γ_{ij}. The one-half factor is necessary in order to make e_{ij} a true tensor. We can write, then, $e_{ii} = \gamma_{ii}$, and $e_{ij} = \frac{1}{2}\gamma_{ij}$ ($i \neq j$).

Appendix A taken from "Basic Equations of Engineering Science" by W. F. Hughes and E. W. Gaylord, copyright 1964, Schaum Publishing Company. Used with permission of McGraw-Hill Book Company. For a more complete listing see this reference.

Cartesian Tensor

$$e_{ji} = e_{ij} = \frac{1}{2}\left(\frac{\partial w_i}{\partial x_j} + \frac{\partial w_j}{\partial x_i}\right)$$

$$-\omega_k = \omega_{ij} = -\omega_{ji} = \frac{1}{2}\left(\frac{\partial w_i}{\partial x_j} - \frac{\partial w_j}{\partial x_i}\right)$$

(A-1)

Cartesian

u, v, and w are the velocities in the x, y, and z directions, respectively.

$$e_{xx} = \frac{\partial u}{\partial x} \qquad e_{xy} = e_{yx} = \frac{1}{2}\left(\frac{\partial u}{\partial y} + \frac{\partial v}{\partial x}\right)$$

$$e_{yy} = \frac{\partial v}{\partial y} \qquad e_{yz} = e_{zy} = \frac{1}{2}\left(\frac{\partial v}{\partial z} + \frac{\partial w}{\partial y}\right)$$

$$e_{zz} = \frac{\partial w}{\partial z} \qquad e_{xz} = e_{zx} = \frac{1}{2}\left(\frac{\partial w}{\partial x} + \frac{\partial u}{\partial z}\right)$$

$$\omega_x = \omega_{zy} = -\omega_{yz} = \frac{1}{2}\left(\frac{\partial w}{\partial y} - \frac{\partial v}{\partial z}\right)$$

$$\omega_y = \omega_{xz} = -\omega_{zx} = \frac{1}{2}\left(\frac{\partial u}{\partial z} - \frac{\partial w}{\partial x}\right)$$

$$\omega_z = \omega_{yx} = -\omega_{xy} = \frac{1}{2}\left(\frac{\partial v}{\partial x} - \frac{\partial u}{\partial y}\right)$$

(A-2)

Cylindrical

v_r, v_θ, and v_z are the velocities in the r, θ, and z directions, respectively.

$$e_{rr} = \frac{\partial v_r}{\partial r} \qquad\qquad e_{r\theta} = e_{\theta r} = \frac{1}{2}\left(\frac{1}{r}\frac{\partial v_r}{\partial \theta} + \frac{\partial v_\theta}{\partial r} - \frac{v_\theta}{r}\right)$$

$$e_{\theta\theta} = \frac{1}{r}\frac{\partial v_\theta}{\partial \theta} + \frac{v_r}{r} \qquad e_{rz} = e_{zr} = \frac{1}{2}\left(\frac{\partial v_r}{\partial z} + \frac{\partial v_z}{\partial r}\right)$$

$$e_{zz} = \frac{\partial v_z}{\partial z} \qquad\qquad e_{\theta z} = e_{z\theta} = \frac{1}{2}\left(\frac{1}{r}\frac{\partial v_z}{\partial \theta} + \frac{\partial v_\theta}{\partial z}\right)$$

$$\omega_r = \omega_{z\theta} = -\omega_{\theta z} = \frac{1}{2}\left(\frac{1}{r}\frac{\partial v_z}{\partial \theta} - \frac{\partial v_\theta}{\partial z}\right)$$

(A-3)

$$\omega_\theta = \omega_{rz} = -\omega_{zr} = \frac{1}{2}\left(\frac{\partial v_r}{\partial z} - \frac{\partial v_z}{\partial r}\right)$$

$$\omega_z = \omega_{\theta r} = -\omega_{r\theta} = \frac{1}{2}\left[\frac{1}{r}\frac{\partial}{\partial r}(rv_\theta) - \frac{1}{r}\frac{\partial v_r}{\partial \theta}\right]$$

(A-3)
(continued)

Spherical

v_r, v_θ, and v_ϕ are the velocities in the r, θ, and ϕ directions, respectively.

$$e_{rr} = \frac{\partial v_r}{\partial r}$$

$$e_{\theta\theta} = \frac{1}{r}\frac{\partial v_\theta}{\partial \theta} + \frac{v_r}{r}$$

$$e_{\phi\phi} = \frac{1}{r\sin\theta}\frac{\partial v_\phi}{\partial \phi} + \frac{v_r}{r} + \frac{v_\theta\cot\theta}{r}$$

$$e_{r\theta} = e_{\theta r} = \frac{1}{2}\left[r\frac{\partial}{\partial r}\left(\frac{v_\theta}{r}\right) + \frac{1}{r}\frac{\partial v_r}{\partial \theta}\right]$$

$$e_{r\phi} = e_{\phi r} = \frac{1}{2}\left[\frac{1}{r\sin\theta}\frac{\partial v_r}{\partial \phi} + r\frac{\partial}{\partial r}\left(\frac{v_\phi}{r}\right)\right]$$

(A-4)

$$e_{\theta\phi} = e_{\phi\theta} = \frac{1}{2}\left[\frac{\sin\theta}{r}\frac{\partial}{\partial \theta}\left(\frac{v_\phi}{\sin\theta}\right) + \frac{1}{r\sin\theta}\frac{\partial v_\theta}{\partial \phi}\right]$$

$$\omega_r = \omega_{\phi\theta} = -\omega_{\theta\phi} = \frac{1}{2r^2\sin\theta}\left[\frac{\partial}{\partial \theta}(rv_\phi\sin\theta) - \frac{\partial}{\partial \phi}(rv_\theta)\right]$$

$$\omega_\theta = \omega_{r\phi} = -\omega_{\phi r} = \frac{1}{2r\sin\theta}\left[\frac{\partial v_r}{\partial \phi} - \frac{\partial}{\partial r}(rv_\phi\sin\theta)\right]$$

$$\omega_\phi = \omega_{\theta r} = -\omega_{r\theta} = \frac{1}{2r}\left[\frac{\partial}{\partial r}(rv_\theta) - \frac{\partial v_r}{\partial \theta}\right]$$

Dilatation

The fluid dilatation ϕ is defined as $e_{11} + e_{22} + e_{33}$ and is exactly equal to the divergence of the velocity. Hence $\phi = \nabla \cdot \mathbf{V}$.

STRESS–STRAIN RATE RELATIONSHIPS

The stress in a fluid may be related to the strain rate by the stress-strain relationships. We confine ourselves to linear relationships here, but more general

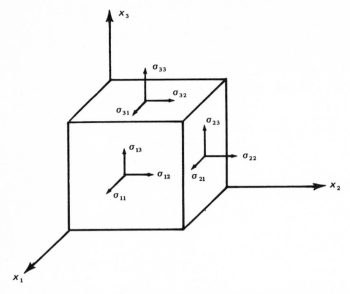

Figure A-1 The stress state on an elemental cube. A surface is denoted by the axis to which it is perpendicular. The stresses shown in the figure are on the positive surfaces. On the opposite or negative surfaces the stresses are in the opposite directions.

relationships, for example, viscoelastic and nonlinear, may be useful in some physical situations.

The components of stress have the following physical significance. The first subscript denotes the normal plane on which the stress acts, and the second subscript denotes the direction in which the stress acts. On the negative normal plane the direction is reversed.

The stress tensor must be symmetrical in order to satisfy equilibrium requirements. Two definitions of the second coefficient of viscosity are used, λ and ζ, both of which are in current usage. For a monatomic gas the second coefficient of viscosity λ is $-\frac{2}{3}\mu$, where μ is the ordinary coefficient of viscosity. ζ is defined as $\zeta = \lambda + \frac{2}{3}\mu$ and is zero for a monatomic gas. The kinematic viscosity ν is μ/ρ, where ρ is the mass density.

In the following, P is pressure, σ_{ij} is the stress tensor, and σ'_{ij} is defined as $\sigma_{ij} + P\delta_{ij}$ and physically is the shear stress tensor plus any normal component due to the second coefficient of viscosity.

δ_{ij} is the Kronecker delta, defined as $\delta_{ij} = 1, i = j; \delta_{ij} = 0, i \neq j$.

The stress-strain rate relationships are as follows:

Cartesian Tensor

w_i is the velocity in the x_i direction. ϕ is dilatation, $\nabla \cdot \mathbf{V}$.

$$\sigma_{ij} = -P\delta_{ij} + \sigma'_{ij} = -P\delta_{ij} + 2\mu e_{ij} + \delta_{ij}\lambda\phi$$

$$= -P\delta_{ij} + \mu\left(\frac{\partial w_i}{\partial x_j} + \frac{\partial w_j}{\partial x_i}\right) + \lambda\delta_{ij}\frac{\partial w_k}{\partial x_k} \qquad \text{(A-5)}$$

$$= -P\delta_{ij} + \mu\left(\frac{\partial w_i}{\partial x_j} + \frac{\partial w_j}{\partial x_i} - \frac{2}{3}\delta_{ij}\frac{\partial w_k}{\partial x_k}\right) + \zeta\delta_{ij}\frac{\partial w_k}{\partial x_k}$$

Cartesian

u, v, and w are the velocities in the x, y, and z directions, respectively.

$$\sigma_{xx} = -P + \sigma'_{xx} = -P + 2\mu e_{xx} + \lambda\nabla \cdot \mathbf{V}$$

$$= -P + 2\mu\frac{\partial u}{\partial x} + \lambda\left(\frac{\partial u}{\partial x} + \frac{\partial v}{\partial y} + \frac{\partial w}{\partial z}\right)$$

$$\sigma_{yy} = -P + \sigma'_{yy} = -P + 2\mu e_{yy} + \lambda\nabla \cdot \mathbf{V}$$

$$= -P + 2\mu\frac{\partial v}{\partial y} + \lambda\left(\frac{\partial u}{\partial x} + \frac{\partial v}{\partial y} + \frac{\partial w}{\partial z}\right)$$

$$\sigma_{zz} = -P + \sigma'_{zz} = -P + 2\mu e_{zz} + \lambda\nabla \cdot \mathbf{V}$$

$$= -P + 2\mu\frac{\partial w}{\partial z} + \lambda\left(\frac{\partial u}{\partial x} + \frac{\partial v}{\partial y} + \frac{\partial w}{\partial z}\right) \qquad \text{(A-6)}$$

$$\sigma_{xy} = \sigma_{yx} = 2\mu e_{xy} = \mu\left(\frac{\partial u}{\partial y} + \frac{\partial v}{\partial x}\right)$$

$$\sigma_{xz} = \sigma_{zx} = 2\mu e_{xz} = \mu\left(\frac{\partial w}{\partial x} + \frac{\partial u}{\partial z}\right)$$

$$\sigma_{yz} = \sigma_{zy} = 2\mu e_{yz} = \mu\left(\frac{\partial v}{\partial z} + \frac{\partial w}{\partial y}\right)$$

Cylindrical

v_r, v_θ, and v_z are the velocities in the r, θ, and z directions, respectively.

$$\sigma_{rr} = -P + \sigma'_{rr} = -P + 2\mu e_{rr} + \lambda\nabla \cdot \mathbf{V}$$

$$= -P + 2\mu\frac{\partial v_r}{\partial r} + \lambda\left[\frac{1}{r}\frac{\partial}{\partial r}(rv_r) + \frac{1}{r}\frac{\partial v_\theta}{\partial \theta} + \frac{\partial v_z}{\partial z}\right] \qquad \text{(A-7)}$$

$$\sigma_{\theta\theta} = -P + \sigma'_{\theta\theta} = -P + 2\mu e_{\theta\theta} + \lambda \nabla \cdot \mathbf{V}$$

$$= -P + 2\mu \left(\frac{1}{r} \frac{\partial v_\theta}{\partial \theta} + \frac{v_r}{r} \right) + \lambda \left[\frac{1}{r} \frac{\partial}{\partial r} (r v_r) + \frac{1}{r} \frac{\partial v_\theta}{\partial \theta} + \frac{\partial v_z}{\partial z} \right]$$

$$\sigma_{zz} = -P + \sigma'_{zz} = -P + 2\mu e_{zz} + \lambda \nabla \cdot \mathbf{V}$$

$$= -P + 2\mu \frac{\partial v_z}{\partial z} + \lambda \left[\frac{1}{r} \frac{\partial}{\partial r} (r v_r) + \frac{1}{r} \frac{\partial v_\theta}{\partial \theta} + \frac{\partial v_z}{\partial z} \right]$$

(A-7)
(continued)

$$\sigma_{r\theta} = \sigma_{\theta r} = 2\mu e_{r\theta} = \mu \left(\frac{1}{r} \frac{\partial v_r}{\partial \theta} + \frac{\partial v_\theta}{\partial r} - \frac{v_\theta}{r} \right)$$

$$\sigma_{rz} = \sigma_{zr} = 2\mu e_{rz} = \mu \left(\frac{\partial v_r}{\partial z} + \frac{\partial v_z}{\partial r} \right)$$

$$\sigma_{\theta z} = \sigma_{z\theta} = 2\mu e_{\theta z} = \mu \left(\frac{1}{r} \frac{\partial v_z}{\partial \theta} + \frac{\partial v_\theta}{\partial z} \right)$$

Spherical

v_r, v_θ, and v_ϕ are the velocities in the r, θ, and ϕ directions, respectively.

$$\sigma_{rr} = -P + \sigma'_{rr} = -P + 2\mu e_{rr} + \lambda \nabla \cdot \mathbf{V}$$

$$= -P + 2\mu \frac{\partial v_r}{\partial r} + \lambda \left[\frac{1}{r^2} \frac{\partial}{\partial r} (r^2 v_r) + \frac{1}{r \sin \theta} \frac{\partial}{\partial \theta} (v_\theta \sin \theta) + \frac{1}{r \sin \theta} \frac{\partial v_\phi}{\partial \phi} \right]$$

$$\sigma_{\theta\theta} = -P + \sigma'_{\theta\theta} = -P + 2\mu e_{\theta\theta} + \lambda \nabla \cdot \mathbf{V}$$

$$= -P + 2\mu \left(\frac{1}{r} \frac{\partial v_\theta}{\partial \theta} + \frac{v_r}{r} \right) + \lambda \left[\frac{1}{r^2} \frac{\partial}{\partial r} (r^2 v_r) + \frac{1}{r \sin \theta} \frac{\partial}{\partial \theta} (v_\theta \sin \theta) + \frac{1}{r \sin \theta} \frac{\partial v_\phi}{\partial \phi} \right]$$

$$\sigma_{\phi\phi} = -P + \sigma'_{\phi\phi} = -P + 2\mu e_{\phi\phi} + \lambda \nabla \cdot \mathbf{V}$$

$$= -P + 2\mu \left(\frac{1}{r \sin \theta} \frac{\partial v_\phi}{\partial \phi} + \frac{v_r}{r} + \frac{v_\theta \cot \theta}{r} \right)$$

$$+ \lambda \left[\frac{1}{r^2} \frac{\partial}{\partial r} (r^2 v_r) + \frac{1}{r \sin \theta} \frac{\partial}{\partial \theta} (v_\theta \sin \theta) + \frac{1}{r \sin \theta} \frac{\partial v_\phi}{\partial \phi} \right]$$

$$\sigma_{r\theta} = \sigma_{\theta r} = 2\mu e_{r\theta} = \mu \left[r \frac{\partial}{\partial r} \left(\frac{v_\theta}{r} \right) + \frac{1}{r} \frac{\partial v_r}{\partial \theta} \right]$$

$$\sigma_{r\phi} = \sigma_{\phi r} = 2\mu e_{r\phi} = \mu \left[\frac{1}{r \sin \theta} \frac{\partial v_r}{\partial \phi} + r \frac{\partial}{\partial r} \left(\frac{v_\phi}{r} \right) \right]$$

(A-8)

$$\sigma_{\theta\phi} = \sigma_{\phi\theta} = 2\mu e_{\theta\phi} = \mu\left[\frac{\sin\theta}{r}\frac{\partial}{\partial\theta}\left(\frac{v_\phi}{\sin\theta}\right) + \frac{1}{r\sin\theta}\frac{\partial v_\theta}{\partial\phi}\right]$$

(A-8)
(*continued*)

NAVIER–STOKES EQUATIONS OF MOTION FOR AN INCOMPRESSIBLE FLUID WITH VISCOSITY CONSTANT

The Navier-Stokes equations may be used with a high degree of accuracy for problems involving viscosity variations if the viscosity gradient is not too large. In most physical problems this assumption is adequate, and the equations may be used in most incompressible flow problems. The following symbols are used:

p = pressure

F = body force density

μ = viscosity

$\dfrac{D}{Dt}$ = material derivative (not the same on a component as on a vector)

Vector

V is the velocity vector.

$$\rho\frac{DV}{Dt} = \rho\left[\frac{\partial V}{\partial t} + (V\cdot\nabla)V\right] = -\nabla p + F + \mu\,\nabla^2 V$$

(A-9)

The term $(V\cdot\nabla)V$ is actually a pseudo-vector expression, and care must be used in its expansion in other than Cartesian coordinates. It is convenient to express this acceleration term in true vector form, and the equation of motion may be written in the alternative form:

$$\rho\left[\frac{\partial V}{\partial t} + \nabla\left(\frac{V^2}{2}\right) - V\times(\nabla\times V)\right] = -\nabla p + F + \mu\,\nabla^2 V$$

(A-10)

Care must be taken in expanding $\nabla^2 V$ and DV/Dt, since the operation on a vector is not the same as the operation on a scalar component. The following vector identity is useful:

$$\nabla^2 V = \nabla(\nabla\cdot V) - \nabla\times(\nabla\times V)$$

Cartesian Tensor

w_i is the velocity in the x_i direction.

$$\rho\left(\frac{\partial w_i}{\partial t} + w_j \frac{\partial w_i}{\partial x_j}\right) = -\frac{\partial p}{\partial x_i} + F_i + \mu \frac{\partial^2 w_i}{\partial x_j \, \partial x_j} \tag{A-11}$$

Cartesian

u, v, and w are the velocities in the x, y, and z directions, respectively. In the following section:

$$\frac{D}{Dt} = \frac{\partial}{\partial t} + u \frac{\partial}{\partial x} + v \frac{\partial}{\partial y} + w \frac{\partial}{\partial z} \qquad \nabla^2 = \frac{\partial^2}{\partial x^2} + \frac{\partial^2}{\partial y^2} + \frac{\partial^2}{\partial z^2}$$

$$\rho \frac{Du}{Dt} = F_x - \frac{\partial p}{\partial x} + \mu \nabla^2 u$$

$$\rho \frac{Dv}{Dt} = F_y - \frac{\partial p}{\partial y} + \mu \nabla^2 v \tag{A-12}$$

$$\rho \frac{Dw}{Dt} = F_z - \frac{\partial p}{\partial z} + \mu \nabla^2 w$$

Written out in full, these become:

$$\rho\left(\frac{\partial u}{\partial t} + u \frac{\partial u}{\partial x} + v \frac{\partial u}{\partial y} + w \frac{\partial u}{\partial z}\right) = -\frac{\partial p}{\partial x} + F_x + \mu\left(\frac{\partial^2 u}{\partial x^2} + \frac{\partial^2 u}{\partial y^2} + \frac{\partial^2 u}{\partial z^2}\right)$$

$$\rho\left(\frac{\partial v}{\partial t} + u \frac{\partial v}{\partial x} + v \frac{\partial v}{\partial y} + w \frac{\partial v}{\partial z}\right) = -\frac{\partial p}{\partial y} + F_y + \mu\left(\frac{\partial^2 v}{\partial x^2} + \frac{\partial^2 v}{\partial y^2} + \frac{\partial^2 v}{\partial z^2}\right) \tag{A-13}$$

$$\rho\left(\frac{\partial w}{\partial t} + u \frac{\partial w}{\partial x} + v \frac{\partial w}{\partial y} + w \frac{\partial w}{\partial z}\right) = -\frac{\partial p}{\partial z} + F_z + \mu\left(\frac{\partial^2 w}{\partial x^2} + \frac{\partial^2 w}{\partial y^2} + \frac{\partial^2 w}{\partial z^2}\right)$$

Cylindrical

v_r, v_θ, and v_z are the velocities in the r, θ, and z directions, respectively. In the following section:

$$\frac{D}{Dt} = \frac{\partial}{\partial t} + v_r \frac{\partial}{\partial r} + \frac{v_\theta}{r} \frac{\partial}{\partial \theta} + v_z \frac{\partial}{\partial z}$$

$$\nabla^2 = \frac{\partial^2}{\partial r^2} + \frac{1}{r} \frac{\partial}{\partial r} + \frac{1}{r^2} \frac{\partial^2}{\partial \theta^2} + \frac{\partial^2}{\partial z^2} \tag{A-14}$$

$$\rho\left(\frac{Dv_r}{Dt} - \frac{v_\theta^2}{r}\right) = F_r - \frac{\partial p}{\partial r} + \mu\left(\nabla^2 v_r - \frac{v_r}{r^2} - \frac{2}{r^2}\frac{\partial v_\theta}{\partial \theta}\right)$$

$$\rho\left(\frac{Dv_\theta}{Dt} + \frac{v_r v_\theta}{r}\right) = F_\theta - \frac{1}{r}\frac{\partial p}{\partial \theta} + \mu\left(\nabla^2 v_\theta + \frac{2}{r^2}\frac{\partial v_r}{\partial \theta} - \frac{v_\theta}{r^2}\right) \qquad \begin{matrix}\text{(A-14)}\\ \textit{(continued)}\end{matrix}$$

$$\rho\frac{Dv_z}{Dt} = F_z - \frac{\partial p}{\partial z} + \mu\,\nabla^2 v_z$$

Written out in full, these become:

$$\rho\left(\frac{\partial v_r}{\partial t} + v_r\frac{\partial v_r}{\partial r} + \frac{v_\theta}{r}\frac{\partial v_r}{\partial \theta} + v_z\frac{\partial v_r}{\partial z} - \frac{v_\theta^2}{r}\right)$$

$$= F_r - \frac{\partial p}{\partial r} + \mu\left(\frac{\partial^2 v_r}{\partial r^2} + \frac{1}{r}\frac{\partial v_r}{\partial r} + \frac{1}{r^2}\frac{\partial^2 v_r}{\partial \theta^2} + \frac{\partial^2 v_r}{\partial z^2} - \frac{v_r}{r^2} - \frac{2}{r^2}\frac{\partial v_\theta}{\partial \theta}\right)$$

$$\rho\left(\frac{\partial v_\theta}{\partial t} + v_r\frac{\partial v_\theta}{\partial r} + \frac{v_\theta}{r}\frac{\partial v_\theta}{\partial \theta} + v_z\frac{\partial v_\theta}{\partial z} + \frac{v_r v_\theta}{r}\right) \qquad \text{(A-15)}$$

$$= F_\theta - \frac{1}{r}\frac{\partial p}{\partial \theta} + \mu\left(\frac{\partial^2 v_\theta}{\partial r^2} + \frac{1}{r}\frac{\partial v_\theta}{\partial r} + \frac{1}{r^2}\frac{\partial^2 v_\theta}{\partial \theta^2} + \frac{\partial^2 v_\theta}{\partial z^2} + \frac{2}{r^2}\frac{\partial v_r}{\partial \theta} - \frac{v_\theta}{r^2}\right)$$

$$\rho\left(\frac{\partial v_z}{\partial t} + v_r\frac{\partial v_z}{\partial r} + \frac{v_\theta}{r}\frac{\partial v_z}{\partial \theta} + v_z\frac{\partial v_z}{\partial z}\right) = F_z - \frac{\partial p}{\partial z} + \mu\left(\frac{\partial^2 v_z}{\partial r^2} + \frac{1}{r}\frac{\partial v_z}{\partial r} + \frac{1}{r^2}\frac{\partial^2 v_z}{\partial \theta^2} + \frac{\partial^2 v_z}{\partial z^2}\right)$$

Spherical

v_r, v_θ, and v_ϕ are the velocities in the r, θ, and ϕ directions, respectively. In the following section:

$$\frac{D}{Dt} = \frac{\partial}{\partial t} + v_r\frac{\partial}{\partial r} + \frac{v_\theta}{r}\frac{\partial}{\partial \theta} + \frac{v_\phi}{r\sin\theta}\frac{\partial}{\partial \phi}$$

$$\nabla^2 = \frac{1}{r^2}\frac{\partial}{\partial r}\left(r^2\frac{\partial}{\partial r}\right) + \frac{1}{r^2\sin\theta}\frac{\partial}{\partial \theta}\left(\sin\theta\frac{\partial}{\partial \theta}\right) + \frac{1}{r^2\sin^2\theta}\frac{\partial^2}{\partial \phi^2}$$

$$\rho\left(\frac{Dv_r}{Dt} - \frac{v_\theta^2 + v_\phi^2}{r}\right)$$

$$= F_r - \frac{\partial p}{\partial r} + \mu\left(\nabla^2 v_r - \frac{2v_r}{r^2} - \frac{2}{r^2}\frac{\partial v_\theta}{\partial \theta} - \frac{2v_\theta\cot\theta}{r^2} - \frac{2}{r^2\sin\theta}\frac{\partial v_\phi}{\partial \phi}\right)$$

$$\rho\left(\frac{Dv_\theta}{Dt} + \frac{v_r v_\theta - v_\phi^2\cot\theta}{r}\right) \qquad \text{(A-16)}$$

$$= F_\theta - \frac{1}{r}\frac{\partial p}{\partial \theta} + \mu\left(\nabla^2 v_\theta + \frac{2}{r^2}\frac{\partial v_r}{\partial \theta} - \frac{v_\theta}{r^2 \sin^2 \theta} - \frac{2\cos\theta}{r^2 \sin^2 \theta}\frac{\partial v_\phi}{\partial \phi}\right)$$

$$\rho\left(\frac{Dv_\phi}{Dt} + \frac{v_\phi v_r}{r} + \frac{v_\theta v_\phi \cot\theta}{r}\right)$$

(A-16)
(continued)

$$= F_\phi - \frac{1}{r\sin\theta}\frac{\partial p}{\partial \phi} + \mu\left(\nabla^2 v_\phi - \frac{v_\theta}{r^2 \sin^2 \theta} + \frac{2}{r^2 \sin^2 \theta}\frac{\partial v_r}{\partial \phi} + \frac{2\cos\theta}{r^2 \sin^2 \theta}\frac{\partial v_\theta}{\partial \phi}\right)$$

Written out in full, these become:

$$\rho\left(\frac{\partial v_r}{\partial t} + v_r\frac{\partial v_r}{\partial r} + \frac{v_\theta}{r}\frac{\partial v_r}{\partial \theta} + \frac{v_\phi}{r\sin\theta}\frac{\partial v_r}{\partial \phi} - \frac{v_\theta^2 + v_\phi^2}{r}\right)$$

$$= F_r - \frac{\partial p}{\partial r} + \mu\left[\frac{1}{r^2}\frac{\partial}{\partial r}\left(r^2\frac{\partial v_r}{\partial r}\right) + \frac{1}{r^2 \sin\theta}\frac{\partial}{\partial \theta}\left(\sin\theta\frac{\partial v_r}{\partial \theta}\right)\right.$$

$$\left. + \frac{1}{r^2 \sin^2\theta}\frac{\partial^2 v_r}{\partial \phi^2} - \frac{2v_r}{r^2} - \frac{2}{r^2}\frac{\partial v_\theta}{\partial \theta} - \frac{2v_\theta\cot\theta}{r^2} - \frac{2}{r^2\sin\theta}\frac{\partial v_\phi}{\partial \phi}\right]$$

$$\rho\left(\frac{\partial v_\theta}{\partial t} + v_r\frac{\partial v_\theta}{\partial r} + \frac{v_\theta}{r}\frac{\partial v_\theta}{\partial \theta} + \frac{v_\phi}{r\sin\theta}\frac{\partial v_\theta}{\partial \phi} + \frac{v_r v_\theta}{r} - \frac{v_\phi^2\cot\theta}{r}\right)$$

$$= F_\theta - \frac{1}{r}\frac{\partial p}{\partial \theta} + \mu\left[\frac{1}{r^2}\frac{\partial}{\partial r}\left(r^2\frac{\partial v_\theta}{\partial r}\right) + \frac{1}{r^2 \sin\theta}\frac{\partial}{\partial \theta}\left(\sin\theta\frac{\partial v_\theta}{\partial \theta}\right)\right. \qquad \text{(A-17)}$$

$$\left. + \frac{1}{r^2 \sin^2\theta}\frac{\partial^2 v_\theta}{\partial \phi^2} + \frac{2}{r^2}\frac{\partial v_r}{\partial \theta} - \frac{v_\theta}{r^2 \sin^2\theta} - \frac{2\cos\theta}{r^2 \sin^2\theta}\frac{\partial v_\phi}{\partial \phi}\right]$$

$$\rho\left(\frac{\partial v_\phi}{\partial t} + v_r\frac{\partial v_\phi}{\partial r} + \frac{v_\theta}{r}\frac{\partial v_\phi}{\partial \theta} + \frac{v_\phi}{r\sin\theta}\frac{\partial v_\phi}{\partial \phi} + \frac{v_\phi v_r}{r} + \frac{v_\theta v_\phi \cot\theta}{r}\right)$$

$$= F_\phi - \frac{1}{r\sin\theta}\frac{\partial p}{\partial \phi} + \mu\left[\frac{1}{r^2}\frac{\partial}{\partial r}\left(r^2\frac{\partial v_\phi}{\partial r}\right) + \frac{1}{r^2 \sin\theta}\frac{\partial}{\partial \theta}\left(\sin\theta\frac{\partial v_\phi}{\partial \theta}\right)\right.$$

$$\left. + \frac{1}{r^2 \sin^2\theta}\frac{\partial^2 v_\phi}{\partial \phi^2} - \frac{v_\phi}{r^2 \sin^2\theta} + \frac{2}{r^2 \sin^2\theta}\frac{\partial v_r}{\partial \phi} + \frac{2\cos\theta}{r^2 \sin^2\theta}\frac{\partial v_\theta}{\partial \phi}\right]$$

THE ENERGY EQUATION

In the following section Q is used to represent internal heat generation density such as that due to Joule heating, chemical, or nuclear reactions, but not by viscous dissipation, which is work accounted for by the mechanical dissipation term Φ. The following symbols are used in this section:

e = specific internal energy (per unit mass)

P = pressure

Q = internal heat generation

$\mathbf{q_r}$ = radiation heat flux vector

T = temperature

κ = thermal conductivity

ρ = mass density

Φ = mechanical or viscous dissipation function

Vector

$\mathbf{V}$ is the velocity vector. D/Dt is the material derivative.

$$\frac{\partial Q}{\partial t} + \Phi + \nabla \cdot (\kappa \, \nabla T) - \nabla \cdot \mathbf{q_r} = \rho \frac{De}{Dt} + P \nabla \cdot \mathbf{V} = \rho \left[\frac{De}{Dt} + P \frac{D}{Dt}\left(\frac{1}{\rho}\right) \right] \quad \text{(A-18)}$$

For constant κ:

$$\frac{\partial Q}{\partial t} + \Phi + \kappa \, \nabla^2 T - \nabla \cdot \mathbf{q_r} = \rho \frac{De}{Dt} + P \nabla \cdot \mathbf{V} = \rho \left[\frac{De}{Dt} + P \frac{D}{Dt}\left(\frac{1}{\rho}\right) \right]$$

For an incompressible fluid:

$$\frac{\partial Q}{\partial t} + \Phi + \nabla \cdot (\kappa \, \nabla T) - \nabla \cdot \mathbf{q_r} = \rho \frac{De}{Dt}$$

For an incompressible fluid with constant κ:

$$\frac{\partial Q}{\partial t} + \Phi + \kappa \, \nabla^2 T - \nabla \cdot \mathbf{q_r} = \rho \frac{De}{Dt}$$

Cartesian Tensor

w_i is the velocity vector in the x_i direction. D/Dt is the material derivative that is written out below.

$$\frac{\partial Q}{\partial t} + \Phi + \frac{\partial}{\partial x_i}\left(\kappa \frac{\partial T}{\partial x_i}\right) - \frac{\partial q_{ri}}{\partial x_i} = \rho \left[\frac{De}{Dt} + P \frac{D}{Dt}\left(\frac{1}{\rho}\right) \right] = \rho \frac{De}{Dt} + P \frac{\partial w_i}{\partial x_i} \quad \text{(A-19)}$$

$$= \rho \left(\frac{\partial e}{\partial t} + w_i \frac{\partial e}{\partial x_i} \right) + P \frac{\partial w_i}{\partial x_i}$$

For constant κ:

$$\frac{\partial Q}{\partial t} + \Phi + \kappa \frac{\partial^2 T}{\partial x_i \, \partial x_i} - \frac{\partial q_{ri}}{\partial x_i} = \rho \left(\frac{\partial e}{\partial t} + w_i \frac{\partial e}{\partial x_i} \right) + P \frac{\partial w_i}{\partial x_i}$$

For an incompressible fluid:

$$\frac{\partial Q}{\partial t} + \Phi + \frac{\partial}{\partial x_i} \left(\kappa \frac{\partial T}{\partial x_i} \right) - \frac{\partial q_{ri}}{\partial x_i} = \rho \left(\frac{\partial e}{\partial t} + w_i \frac{\partial e}{\partial x_i} \right)$$

For an incompressible fluid with constant κ:

$$\frac{\partial Q}{\partial t} + \Phi + \kappa \frac{\partial^2 T}{\partial x_i \, \partial x_i} - \frac{\partial q_{ri}}{\partial x_i} = \rho \left(\frac{\partial e}{\partial t} + w_i \frac{\partial e}{\partial x_i} \right)$$

Cartesian

u, v, and w are the velocities in the x, y, and z directions, respectively. In the following section:

$$\frac{D}{Dt} = \frac{\partial}{\partial t} + u \frac{\partial}{\partial x} + v \frac{\partial}{\partial y} + w \frac{\partial}{\partial z}$$

$$\nabla \cdot \mathbf{V} = \frac{\partial u}{\partial x} + \frac{\partial v}{\partial y} + \frac{\partial w}{\partial z} \qquad \text{(A-20)}$$

$$\nabla^2 = \frac{\partial^2}{\partial x^2} + \frac{\partial^2}{\partial y^2} + \frac{\partial^2}{\partial z^2}$$

The general equation is

$$\frac{\partial Q}{\partial t} + \Phi + \frac{\partial}{\partial x} \left(\kappa \frac{\partial T}{\partial x} \right) + \frac{\partial}{\partial y} \left(\kappa \frac{\partial T}{\partial y} \right) + \frac{\partial}{\partial z} \left(\kappa \frac{\partial T}{\partial z} \right) - \nabla \cdot \mathbf{q_r} = \rho \left[\frac{De}{Dt} + P \frac{D}{Dt} \left(\frac{1}{\rho} \right) \right]$$

$$= \rho \frac{De}{Dt} + P \nabla \cdot \mathbf{V}$$

For constant κ:

$$\frac{\partial Q}{\partial t} + \Phi + \kappa \nabla^2 T - \nabla \cdot \mathbf{q_r} = \rho \left[\frac{De}{Dt} + P \frac{D}{Dt} \left(\frac{1}{\rho} \right) \right]$$

$$= \rho \frac{De}{Dt} + P \nabla \cdot \mathbf{V}$$

For an incompressible fluid:

$$\frac{\partial Q}{\partial t} + \Phi + \frac{\partial}{\partial x}\left(\kappa\,\frac{\partial T}{\partial x}\right) + \frac{\partial}{\partial y}\left(\kappa\,\frac{\partial T}{\partial y}\right) + \frac{\partial}{\partial z}\left(\kappa\,\frac{\partial T}{\partial z}\right) - \nabla\cdot\mathbf{q_r} = \rho\,\frac{De}{Dt}$$

For an incompressible fluid with constant κ:

$$\frac{\partial Q}{\partial t} + \Phi + \kappa\,\nabla^2 T - \nabla\cdot\mathbf{q_r} = \rho\,\frac{De}{Dt}$$

Cylindrical

v_r, v_θ, and v_z are the velocities in the r, θ, and z directions, respectively. In the following section:

$$\frac{D}{Dt} = \frac{\partial}{\partial t} + v_r\,\frac{\partial}{\partial r} + \frac{v_\theta}{r}\,\frac{\partial}{\partial\theta} + v_z\,\frac{\partial}{\partial z}$$

$$\nabla\cdot\mathbf{V} = \frac{1}{r}\,\frac{\partial}{\partial r}(rv_r) + \frac{1}{r}\,\frac{\partial v_\theta}{\partial\theta} + \frac{\partial v_z}{\partial z} \qquad (A\text{-}21)$$

$$\nabla^2 = \frac{\partial^2}{\partial r^2} + \frac{1}{r}\,\frac{\partial}{\partial r} + \frac{1}{r^2}\,\frac{\partial^2}{\partial\theta^2} + \frac{\partial^2}{\partial z^2}$$

The general equation is

$$\frac{\partial Q}{\partial t} + \Phi + \frac{1}{r}\,\frac{\partial}{\partial r}\left(r\kappa\,\frac{\partial T}{\partial r}\right) + \frac{1}{r^2}\,\frac{\partial}{\partial\theta}\left(\kappa\,\frac{\partial T}{\partial\theta}\right) + \frac{\partial}{\partial z}\left(\kappa\,\frac{\partial T}{\partial z}\right) - \nabla\cdot\mathbf{q_r} = \rho\,\frac{De}{Dt} + P\nabla\cdot\mathbf{V}$$

$$= \rho\left[\frac{De}{Dt} + P\,\frac{D}{Dt}\left(\frac{1}{\rho}\right)\right]$$

For constant κ:

$$\frac{\partial Q}{\partial t} + \Phi + \kappa\,\nabla^2 T - \nabla\cdot\mathbf{q_r} = \rho\,\frac{De}{Dt} + P\nabla\cdot\mathbf{V}$$

$$= \rho\left[\frac{De}{Dt} + P\,\frac{D}{Dt}\left(\frac{1}{\rho}\right)\right]$$

For an incompressible fluid:

$$\frac{\partial Q}{\partial t} + \Phi + \frac{1}{r}\,\frac{\partial}{\partial r}\left(r\kappa\,\frac{\partial T}{\partial r}\right) + \frac{1}{r^2}\,\frac{\partial}{\partial\theta}\left(\kappa\,\frac{\partial T}{\partial\theta}\right) + \frac{\partial}{\partial z}\left(\kappa\,\frac{\partial T}{\partial z}\right) - \nabla\cdot\mathbf{q_r} = \rho\,\frac{De}{Dt}$$

For an incompressible fluid with constant κ:

$$\frac{\partial Q}{\partial t} + \Phi + \kappa \nabla^2 T - \nabla \cdot \mathbf{q_r} = \rho \frac{De}{Dt}$$

Spherical

v_r, v_θ, and v_ϕ are the velocities in the r, θ, and ϕ directions, respectively. In the following section:

$$\frac{D}{Dt} = \frac{\partial}{\partial t} + v_r \frac{\partial}{\partial r} + \frac{v_\theta}{r} \frac{\partial}{\partial \theta} + \frac{v_\phi}{r \sin \theta} \frac{\partial}{\partial \phi}$$

$$\nabla \cdot \mathbf{V} = \frac{1}{r^2} \frac{\partial}{\partial r}(r^2 v_r) + \frac{1}{r \sin \theta} \frac{\partial}{\partial \theta}(v_\theta \sin \theta) + \frac{1}{r \sin \theta} \frac{\partial v_\phi}{\partial \phi} \qquad \text{(A-22)}$$

$$\nabla^2 = \frac{1}{r^2} \frac{\partial}{\partial r}\left(r^2 \frac{\partial}{\partial r}\right) + \frac{1}{r^2 \sin \theta} \frac{\partial}{\partial \theta}\left(\sin \theta \frac{\partial}{\partial \theta}\right) + \frac{1}{r^2 \sin^2 \theta} \frac{\partial^2}{\partial \phi^2}$$

The general equation is

$$\frac{\partial Q}{\partial t} + \Phi + \frac{1}{r^2} \frac{\partial}{\partial r}\left(r^2 \kappa \frac{\partial T}{\partial r}\right) + \frac{1}{r^2 \sin \theta} \frac{\partial}{\partial \theta}\left(\kappa \sin \theta \frac{\partial T}{\partial \theta}\right) + \frac{1}{r^2 \sin^2 \theta} \frac{\partial}{\partial \phi}\left(\kappa \frac{\partial T}{\partial \phi}\right) - \nabla \cdot \mathbf{q_r}$$

$$= \rho \frac{De}{Dt} + P \nabla \cdot \mathbf{V} = \rho \left[\frac{De}{Dt} + P \frac{D}{Dt}\left(\frac{1}{\rho}\right)\right]$$

For constant κ:

$$\frac{\partial Q}{\partial t} + \Phi + k \nabla^2 T - \nabla \cdot \mathbf{q_r} = \rho \frac{De}{Dt} + P \nabla \cdot \mathbf{V}$$

$$= \rho \left[\frac{De}{Dt} + P \frac{D}{Dt}\left(\frac{1}{\rho}\right)\right]$$

For an incompressible fluid:

$$\frac{\partial Q}{\partial t} + \Phi + \frac{1}{r^2} \frac{\partial}{\partial r}\left(r^2 \kappa \frac{\partial T}{\partial r}\right) + \frac{1}{r^2 \sin \theta} \frac{\partial}{\partial \theta}\left(\kappa \sin \theta \frac{\partial T}{\partial \theta}\right)$$

$$+ \frac{1}{r^2 \sin^2 \theta} \frac{\partial}{\partial \phi}\left(\kappa \frac{\partial T}{\partial \phi}\right) - \nabla \cdot \mathbf{q_r} = \rho \frac{De}{Dt}$$

For an incompressible fluid with constant κ:

$$\frac{\partial Q}{\partial t} + \Phi + \kappa \nabla^2 T - \nabla \cdot \mathbf{q_r} = \rho \frac{De}{Dt}$$

ENERGY EQUATION IN TERMS OF ENTHALPY

The energy equation can, generally, be expressed in terms of enthalpy instead of internal energy. The specific enthalpy (per unit mass) is denoted by h. The material derivative D/Dt has been written out in the previous section for various coordinate systems and is not repeated here.

Vector

$$\rho \frac{Dh}{Dt} = \frac{DP}{Dt} + \frac{\partial Q}{\partial t} + \Phi + \nabla \cdot (\kappa \nabla T) - \nabla \cdot \mathbf{q_r}$$

For constant κ: (A-23)

$$\rho \frac{Dh}{Dt} = \frac{DP}{Dt} + \frac{\partial Q}{\partial t} + \Phi + \kappa \nabla^2 T - \nabla \cdot \mathbf{q_r}$$

Cartesian Tensor

$$\rho \frac{Dh}{Dt} = \frac{DP}{Dt} + \frac{\partial Q}{\partial t} + \Phi + \frac{\partial}{\partial x_i}\left(\kappa \frac{\partial T}{\partial x_i}\right) - \frac{\partial q_{ri}}{\partial x_i}$$

For constant κ: (A-24)

$$\rho \frac{Dh}{Dt} = \frac{DP}{Dt} + \frac{\partial Q}{\partial t} + \Phi + \kappa \frac{\partial^2 T}{\partial x_i \partial x_i} - \frac{\partial q_{ri}}{\partial x_i}$$

Cartesian

$$\rho \frac{Dh}{Dt} = \frac{DP}{Dt} + \frac{\partial Q}{\partial t} + \Phi + \frac{\partial}{\partial x}\left(\kappa \frac{\partial T}{\partial x}\right) + \frac{\partial}{\partial y}\left(\kappa \frac{\partial T}{\partial y}\right) + \frac{\partial}{\partial z}\left(\kappa \frac{\partial T}{\partial z}\right) - \nabla \mathbf{q_r}$$

For constant κ: (A-25)

$$\rho \frac{Dh}{Dt} = \frac{DP}{Dt} + \frac{\partial Q}{\partial t} + \Phi + \kappa \nabla^2 T - \nabla \cdot \mathbf{q_r}$$

Cylindrical

$$\rho \frac{Dh}{Dt} = \frac{DP}{Dt} + \frac{\partial Q}{\partial t} + \Phi + \frac{1}{r}\frac{\partial}{\partial r}\left(r\kappa \frac{\partial T}{\partial r}\right) + \frac{1}{r^2}\frac{\partial}{\partial \theta}\left(\kappa \frac{\partial T}{\partial \theta}\right) + \frac{\partial}{\partial z}\left(\kappa \frac{\partial T}{\partial z}\right) - \nabla \cdot \mathbf{q_r}$$

For constant κ: (A-26)

$$\rho \frac{Dh}{Dt} = \frac{DP}{Dt} + \frac{\partial Q}{\partial t} + \Phi + \kappa \nabla^2 T - \nabla \cdot \mathbf{q_r}$$

Spherical

$$\rho \frac{Dh}{Dt} = \frac{DP}{Dt} + \frac{\partial Q}{\partial t} + \Phi + \frac{1}{r^2}\frac{\partial}{\partial r}\left(r^2\kappa \frac{\partial T}{\partial r}\right) + \frac{1}{r^2 \sin \theta}\frac{\partial}{\partial \theta}\left(\kappa \sin \theta \frac{\partial T}{\partial \theta}\right)$$

$$+ \frac{1}{r^2 \sin^2 \theta}\frac{\partial}{\partial \phi}\left(\kappa \frac{\partial T}{\partial \phi}\right) - \nabla \cdot \mathbf{q_r}$$

For constant κ: (A-27)

$$\rho \frac{Dh}{Dt} = \frac{DP}{Dt} + \frac{\partial Q}{\partial t} + \Phi + \kappa \nabla^2 T - \nabla \cdot \mathbf{q_r}$$

Perfect Gas

For a perfect gas the material derivative of enthalpy or internal energy that occurs in the previous sections can be written as

$$\frac{Dh}{Dt} = c_p \frac{DT}{Dt} \quad \text{and} \quad \frac{De}{Dt} = c_v \frac{DT}{Dt}$$

THE DISSIPATION FUNCTION

The mechanical or viscous dissipation function Φ is defined in generalized orthogonal coordinates as (in terms of the strain rate tensor e_{ij})

$$\Phi = \mu[2(e_{11}^2 + e_{22}^2 + e_{33}^2) + (2e_{23})^2 + (2e_{31})^2 + (2e_{12})^2] + \lambda(e_{11} + e_{22} + e_{33})^2$$

(A-28)

(In some texts, the definition of Φ differs by a factor of μ from the one defined here.) λ is the second coefficient of viscosity defined in the previous section.

Cartesian Tensor in Terms of the Stress Tensor

$$\Phi = \sigma'_{ij} \frac{\partial w_i}{\partial x_j} \tag{A-29}$$

Cartesian

$$\Phi = 2\mu \left[\left(\frac{\partial u}{\partial x}\right)^2 + \left(\frac{\partial v}{\partial y}\right)^2 + \left(\frac{\partial w}{\partial z}\right)^2 + \frac{1}{2}\left(\frac{\partial u}{\partial y} + \frac{\partial v}{\partial x}\right)^2 + \frac{1}{2}\left(\frac{\partial v}{\partial z} + \frac{\partial w}{\partial y}\right)^2 \right.$$

$$\left. + \frac{1}{2}\left(\frac{\partial w}{\partial x} + \frac{\partial u}{\partial z}\right)^2 \right] + \lambda\left(\frac{\partial u}{\partial x} + \frac{\partial v}{\partial y} + \frac{\partial w}{\partial z}\right)^2 \tag{A-30}$$

Cylindrical

$$\Phi = \mu \left\{ 2\left[\left(\frac{\partial v_r}{\partial r}\right)^2 + \left(\frac{1}{r}\frac{\partial v_\theta}{\partial \theta} + \frac{v_r}{r}\right)^2 + \left(\frac{\partial v_z}{\partial z}\right)^2 \right] + \left(\frac{1}{r}\frac{\partial v_z}{\partial \theta} + \frac{\partial v_\theta}{\partial z}\right)^2 \right.$$

$$\left. + \left(\frac{\partial v_r}{\partial z} + \frac{\partial v_z}{\partial r}\right)^2 + \left(\frac{1}{r}\frac{\partial v_r}{\partial \theta} + \frac{\partial v_\theta}{\partial r} - \frac{v_\theta}{r}\right)^2 \right\} \tag{A-31}$$

$$+ \lambda\left(\frac{\partial v_r}{\partial r} + \frac{1}{r}\frac{\partial v_\theta}{\partial \theta} + \frac{v_r}{r} + \frac{\partial v_z}{\partial z}\right)^2$$

Spherical

$$\Phi = \mu \left\{ 2\left[\left(\frac{\partial v_r}{\partial r}\right)^2 + \left(\frac{1}{r}\frac{\partial v_\theta}{\partial \theta} + \frac{v_r}{r}\right)^2 + \left(\frac{1}{r\sin\theta}\frac{\partial v_\phi}{\partial \phi} + \frac{v_r}{r} + \frac{v_\theta \cot\theta}{r}\right)^2 \right] \right.$$

$$+ \left[\frac{1}{r\sin\theta}\frac{\partial v_\theta}{\partial \phi} + \frac{\sin\theta}{r}\frac{\partial}{\partial \theta}\left(\frac{v_\phi}{\sin\theta}\right)\right]^2$$

$$\left. + \left[\frac{1}{r\sin\theta}\frac{\partial v_r}{\partial \phi} + r\frac{\partial}{\partial r}\left(\frac{v_\phi}{r}\right)\right]^2 + \left[r\frac{\partial}{\partial r}\left(\frac{v_\theta}{r}\right) + \frac{1}{r}\frac{\partial v_r}{\partial \theta}\right]^2 \right\} \tag{A-32}$$

$$+ \lambda\left(\frac{\partial v_r}{\partial r} + \frac{1}{r}\frac{\partial v_\theta}{\partial \theta} + \frac{2v_r}{r} + \frac{1}{r\sin\theta}\frac{\partial v_\phi}{\partial \phi} + \frac{v_\theta \cot\theta}{r}\right)^2$$

SOME USEFUL VECTOR OPERATIONS

The operations D/Dt and ∇^2 listed below are for operations *on a scalar*. These are not the same as operation on a vector except in Cartesian coordinates.

Cartesian on a Scalar

$$\frac{D}{Dt} = \frac{\partial}{\partial t} + u\frac{\partial}{\partial x} + v\frac{\partial}{\partial y} + w\frac{\partial}{\partial z}$$

$$\nabla \cdot \mathbf{V} = \frac{\partial u}{\partial x} + \frac{\partial v}{\partial y} + \frac{\partial w}{\partial z} \tag{A-33}$$

$$\nabla^2 = \frac{\partial^2}{\partial x^2} + \frac{\partial^2}{\partial y^2} + \frac{\partial^2}{\partial z^2}$$

Cylindrical on a Scalar

$$\frac{D}{Dt} = \frac{\partial}{\partial t} + v_r\frac{\partial}{\partial r} + \frac{v_\theta}{r}\frac{\partial}{\partial \theta} + v_z\frac{\partial}{\partial z}$$

$$\nabla \cdot \mathbf{V} = \frac{1}{r}\frac{\partial}{\partial r}(rv_r) + \frac{1}{r}\frac{\partial v_\theta}{\partial \theta} + \frac{\partial v_z}{\partial z} \tag{A-34}$$

$$\nabla^2 = \frac{\partial^2}{\partial r^2} + \frac{1}{r}\frac{\partial}{\partial r} + \frac{1}{r^2}\frac{\partial^2}{\partial \theta^2} + \frac{\partial^2}{\partial z^2}$$

Spherical on a Scalar

$$\frac{D}{Dt} = \frac{\partial}{\partial t} + v_r\frac{\partial}{\partial r} + \frac{v_\theta}{r}\frac{\partial}{\partial \theta} + \frac{v_\phi}{r\sin\theta}\frac{\partial}{\partial \phi}$$

$$\nabla \cdot \mathbf{V} = \frac{1}{r^2}\frac{\partial}{\partial r}(r^2 v_r) + \frac{1}{r\sin\theta}\frac{\partial}{\partial \theta}(v_\theta \sin\theta) + \frac{1}{r\sin\theta}\frac{\partial v_\phi}{\partial \phi} \tag{A-35}$$

$$\nabla^2 = \frac{1}{r^2}\frac{\partial}{\partial r}\left(r^2\frac{\partial}{\partial r}\right) + \frac{1}{r^2\sin\theta}\frac{\partial}{\partial \theta}\left(\sin\theta\frac{\partial}{\partial \theta}\right) + \frac{1}{r^2\sin^2\theta}\frac{\partial^2}{\partial \phi^2}$$

The operations D/Dt and ∇^2 listed below are written out in component form for operation *on a vector* **V**.

Cartesian on a Vector (components of V are u, v, w)

$$\left.\frac{D\mathbf{V}}{Dt}\right|_x = \frac{\partial u}{\partial t} + u\frac{\partial u}{\partial x} + v\frac{\partial u}{\partial y} + w\frac{\partial u}{\partial z} \tag{A-36}$$

$$\frac{D\mathbf{V}}{Dt}\bigg|_y = \frac{\partial v}{\partial t} + u\frac{\partial v}{\partial x} + v\frac{\partial v}{\partial y} + w\frac{\partial v}{\partial z}$$

$$\frac{D\mathbf{V}}{Dt}\bigg|_z = \frac{\partial w}{\partial t} + u\frac{\partial w}{\partial x} + v\frac{\partial w}{\partial y} + w\frac{\partial w}{\partial z}$$

$$\nabla^2\mathbf{V}\big|_x = \frac{\partial^2 u}{\partial x^2} + \frac{\partial^2 u}{\partial y^2} + \frac{\partial^2 u}{\partial z^2}$$

$$\nabla^2\mathbf{V}\big|_y = \frac{\partial^2 v}{\partial x^2} + \frac{\partial^2 v}{\partial y^2} + \frac{\partial^2 v}{\partial z^2}$$

(A-36)
(*continued*)

$$\nabla^2\mathbf{V}\big|_z = \frac{\partial^2 w}{\partial x^2} + \frac{\partial^2 w}{\partial y^2} + \frac{\partial^2 w}{\partial z^2}$$

Cylindrical on a Vector (components of V are v_r, v_θ, v_z)

$$\frac{D\mathbf{V}}{Dt}\bigg|_r = \frac{\partial v_r}{\partial t} + v_r\frac{\partial v_r}{\partial r} + \frac{v_\theta}{r}\frac{\partial v_r}{\partial \theta} + v_z\frac{\partial v_r}{\partial z} - \frac{v_\theta^2}{r} = \frac{Dv_r}{Dt} - \frac{v_\theta^2}{r}$$

$$\frac{D\mathbf{V}}{Dt}\bigg|_\theta = \frac{\partial v_\theta}{\partial t} + v_r\frac{\partial v_\theta}{\partial r} + \frac{v_\theta}{r}\frac{\partial v_\theta}{\partial \theta} + v_z\frac{\partial v_\theta}{\partial z} + \frac{v_r v_\theta}{r} = \frac{Dv_\theta}{Dt} + \frac{v_r v_\theta}{r}$$

$$\frac{D\mathbf{V}}{Dt}\bigg|_z = \frac{\partial v_z}{\partial t} + v_r\frac{\partial v_z}{\partial r} + \frac{v_\theta}{r}\frac{\partial v_z}{\partial \theta} + v_z\frac{\partial v_z}{\partial z} = \frac{Dv_z}{Dt}$$

(A-37)

$$\nabla^2\mathbf{V}\big|_r = \nabla^2 v_r - \frac{v_r}{r^2} - \frac{2}{r^2}\frac{\partial v_\theta}{\partial \theta} = \frac{\partial^2 v_r}{\partial r^2} + \frac{1}{r}\frac{\partial v_r}{\partial r} + \frac{1}{r^2}\frac{\partial^2 v_r}{\partial \theta^2} + \frac{\partial^2 v_r}{\partial z^2} - \frac{v_r}{r^2} - \frac{2}{r^2}\frac{\partial v_\theta}{\partial \theta}$$

$$\nabla^2\mathbf{V}\big|_\theta = \nabla^2 v_\theta + \frac{2}{r^2}\frac{\partial v_r}{\partial \theta} - \frac{v_\theta}{r^2} = \frac{\partial^2 v_\theta}{\partial r^2} + \frac{1}{r}\frac{\partial v_\theta}{\partial r} + \frac{1}{r^2}\frac{\partial^2 v_\theta}{\partial \theta^2} + \frac{\partial^2 v_\theta}{\partial z^2} + \frac{2}{r^2}\frac{\partial v_r}{\partial \theta} - \frac{v_\theta}{r^2}$$

$$\nabla^2\mathbf{V}\big|_z = \nabla^2 v_z = \frac{\partial^2 v_z}{\partial r^2} + \frac{1}{r}\frac{\partial v_z}{\partial r} + \frac{1}{r^2}\frac{\partial^2 v_z}{\partial \theta^2} + \frac{\partial^2 v_z}{\partial z^2}$$

Spherical on a Vector (components of V are v_r, v_θ, v_ϕ)

$$\frac{D\mathbf{V}}{Dt}\bigg|_r = \frac{Dv_r}{Dt} - \frac{v_\theta^2 + v_\phi^2}{r} = \frac{\partial v_r}{\partial t} + v_r\frac{\partial v_r}{\partial r} + \frac{v_\theta}{r}\frac{\partial v_r}{\partial \theta} + \frac{v_\phi}{r\sin\theta}\frac{\partial v_r}{\partial \phi} - \frac{v_\theta^2 + v_\phi^2}{r}$$

$$\frac{D\mathbf{V}}{Dt}\bigg|_\theta = \frac{Dv_\theta}{Dt} + \frac{v_r v_\theta - v_\phi^2\cot\theta}{r}$$

(A-38)

$$= \frac{\partial v_\theta}{\partial t} + v_r\frac{\partial v_\theta}{\partial r} + \frac{v_\theta}{r}\frac{\partial v_\theta}{\partial \theta} + \frac{v_\phi}{r\sin\theta}\frac{\partial v_\theta}{\partial \phi} + \frac{v_r v_\theta - v_\phi^2\cot\theta}{r}$$

$$\frac{D\mathbf{V}}{Dt}\bigg|_\phi = \frac{Dv_\phi}{Dt} + \frac{v_\phi v_r + v_\theta v_\phi \cot\theta}{r}$$

$$= \frac{\partial v_\phi}{\partial t} + v_r \frac{\partial v_\phi}{\partial r} + \frac{v_\theta}{r}\frac{\partial v_\phi}{\partial\theta} + \frac{v_\phi}{r\sin\theta}\frac{\partial v_\phi}{\partial\phi} + \frac{v_\phi v_r + v_\theta v_\phi \cot\theta}{r}$$

$$\nabla^2\mathbf{V}\big|_r = \nabla^2 v_r - \frac{2v_r}{r^2} - \frac{2}{r^2}\frac{\partial v_\theta}{\partial\theta} - \frac{2v_\theta \cot\theta}{r^2} - \frac{2}{r^2\sin\theta}\frac{\partial v_\phi}{\partial\phi}$$

$$= \frac{1}{r^2}\frac{\partial}{\partial r}\left(r^2\frac{\partial v_r}{\partial r}\right) + \frac{1}{r^2\sin\theta}\frac{\partial}{\partial\theta}\left(\sin\theta\frac{\partial v_r}{\partial\theta}\right) + \frac{1}{r^2\sin^2\theta}\frac{\partial^2 v_r}{\partial\phi^2}$$

$$- \frac{2v_r}{r^2} - \frac{2}{r^2}\frac{\partial v_\theta}{\partial\theta} - \frac{2v_\theta \cot\theta}{r^2} - \frac{2}{r^2\sin\theta}\frac{\partial v_\phi}{\partial\phi}$$

$$\nabla^2\mathbf{V}\big|_\theta = \nabla^2 v_\theta + \frac{2}{r^2}\frac{\partial v_r}{\partial\theta} - \frac{v_\theta}{r^2\sin^2\theta} - \frac{2\cos\theta}{r^2\sin^2\theta}\frac{\partial v_\phi}{\partial\phi} \qquad\text{(A-38)}$$

$$\qquad\qquad\qquad\qquad\qquad\qquad\qquad\qquad\qquad\qquad\qquad\quad (\textit{continued})$$

$$= \frac{1}{r^2}\frac{\partial}{\partial r}\left(r^2\frac{\partial v_\theta}{\partial r}\right) + \frac{1}{r^2\sin\theta}\frac{\partial}{\partial\theta}\left(\sin\theta\frac{\partial v_\theta}{\partial\theta}\right) + \frac{1}{r^2\sin^2\theta}\frac{\partial^2 v_\theta}{\partial\phi^2}$$

$$+ \frac{2}{r^2}\frac{\partial v_r}{\partial\theta} - \frac{v_\theta}{r^2\sin^2\theta} - \frac{2\cos\theta}{r^2\sin^2\theta}\frac{\partial v_\phi}{\partial\phi}$$

$$\nabla^2\mathbf{V}\big|_\phi = \nabla^2 v_\phi - \frac{v_\phi}{r^2\sin^2\theta} + \frac{2}{r^2\sin^2\theta}\frac{\partial v_r}{\partial\phi} + \frac{2\cos\theta}{r^2\sin^2\theta}\frac{\partial v_\theta}{\partial\phi}$$

$$= \frac{1}{r^2}\frac{\partial}{\partial r}\left(r^2\frac{\partial v_\phi}{\partial r}\right) + \frac{1}{r^2\sin\theta}\frac{\partial}{\partial\theta}\left(\sin\theta\frac{\partial v_\phi}{\partial\theta}\right) + \frac{1}{r^2\sin^2\theta}\frac{\partial^2 v_\phi}{\partial\phi^2}$$

$$- \frac{v_\phi}{r^2\sin^2\theta} + \frac{2}{r^2\sin^2\theta}\frac{\partial v_r}{\partial\phi} + \frac{2\cos\theta}{r^2\sin^2\theta}\frac{\partial v_\theta}{\partial\phi}$$

VECTOR IDENTITIES

A is a vector defined as

$$\mathbf{A} = \mathbf{e}_1 A_1 + \mathbf{e}_2 A_2 + \mathbf{e}_3 A_3 \qquad\text{(A-39)}$$

where $\mathbf{e}_1$, $\mathbf{e}_2$, and $\mathbf{e}_3$ are unit vectors in the coordinate directions and A_1, A_2, and A_3 are the components of the vector. A scalar is denoted as Φ. ∇ is the operator "del."

$$\mathbf{A}\cdot\mathbf{B} = A_1 B_1 + A_2 B_2 + A_3 B_3 \qquad\text{(A-40)}$$

$$\mathbf{A} \cdot \mathbf{B} = \mathbf{B} \cdot \mathbf{A} \tag{A-41}$$

$$\mathbf{A} \cdot (\mathbf{B} + \mathbf{C}) = \mathbf{A} \cdot \mathbf{B} + \mathbf{A} \cdot \mathbf{C} \tag{A-42}$$

$$\mathbf{A} \times \mathbf{B} = -\mathbf{B} \times \mathbf{A} = \begin{vmatrix} e_1 & e_2 & e_3 \\ A_1 & A_2 & A_3 \\ B_1 & B_2 & B_3 \end{vmatrix} \tag{A-43}$$

$$(\mathbf{A} + \mathbf{B}) \times \mathbf{C} = (\mathbf{A} \times \mathbf{C}) + (\mathbf{B} \times \mathbf{C}) \tag{A-44}$$

$$\mathbf{A} \times (\mathbf{B} + \mathbf{C}) = (\mathbf{A} \times \mathbf{B}) + (\mathbf{A} \times \mathbf{C}) \tag{A-45}$$

$$\mathbf{A} \times (\mathbf{B} \times \mathbf{C}) = \mathbf{B}(\mathbf{A} \cdot \mathbf{C}) - \mathbf{C}(\mathbf{A} \cdot \mathbf{B}) \tag{A-46}$$

$$\mathbf{A} \cdot (\mathbf{B} \times \mathbf{C}) = (\mathbf{A} \times \mathbf{B}) \cdot \mathbf{C} = \mathbf{B} \cdot (\mathbf{C} \times \mathbf{A}) = \begin{vmatrix} A_1 & A_2 & A_3 \\ B_1 & B_2 & B_3 \\ C_1 & C_2 & C_3 \end{vmatrix} \tag{A-47}$$

$$(\mathbf{A} \times \mathbf{B}) \cdot (\mathbf{C} \times \mathbf{D}) = (\mathbf{A} \cdot \mathbf{C})(\mathbf{B} \cdot \mathbf{D}) - (\mathbf{A} \cdot \mathbf{D})(\mathbf{B} \cdot \mathbf{C}) \tag{A-48}$$

$$(\mathbf{A} \times \mathbf{B}) \times (\mathbf{C} \times \mathbf{D}) = \mathbf{B}[\mathbf{A} \cdot (\mathbf{C} \times \mathbf{D})] - \mathbf{A}[\mathbf{B} \cdot (\mathbf{C} \times \mathbf{D})] \tag{A-49}$$
$$= \mathbf{C}[\mathbf{A} \cdot (\mathbf{B} \times \mathbf{D})] - \mathbf{D}[\mathbf{A} \cdot (\mathbf{B} \times \mathbf{C})]$$

$$\nabla^2 \Phi = \nabla \cdot \nabla \Phi \tag{A-50}$$

$$\nabla^2 \mathbf{A} = (\nabla \cdot \nabla)\mathbf{A} \tag{A-51}$$

$$\nabla \cdot \nabla \times \mathbf{A} = 0 \tag{A-52}$$

$$\nabla \times \nabla \Phi = 0 \tag{A-53}$$

$$\nabla \times (\nabla \times \mathbf{A}) = \nabla(\nabla \cdot \mathbf{A}) - \nabla^2 \mathbf{A} \tag{A-54}$$

$$(\mathbf{A} \cdot \nabla)\mathbf{A} = \nabla\left(\frac{|\mathbf{A}|^2}{2}\right) - \mathbf{A} \times (\nabla \times \mathbf{A}) \tag{A-55}$$

$$\nabla \times (\mathbf{A} \times \mathbf{B}) = (\mathbf{B} \cdot \nabla)\mathbf{A} - \mathbf{B}(\nabla \cdot \mathbf{A}) - (\mathbf{A} \cdot \nabla)\mathbf{B} + \mathbf{A}(\nabla \cdot \mathbf{B}) \tag{A-56}$$

$$\nabla \cdot (\mathbf{A} \times \mathbf{B}) = \mathbf{B} \cdot \nabla \times \mathbf{A} - \mathbf{A} \cdot \nabla \times \mathbf{B} \tag{A-57}$$

$$\nabla(\mathbf{A} \cdot \mathbf{B}) = (\mathbf{B} \cdot \nabla)\mathbf{A} + (\mathbf{A} \cdot \nabla)\mathbf{B} + \mathbf{B} \times (\nabla \times \mathbf{A}) + \mathbf{A} \times (\nabla \times \mathbf{B}) \tag{A-58}$$

B

PROPERTIES OF FLUIDS

Table B-1 Properties of Water at Atmospheric Pressure

Temperature		Density			Viscosity			Kinematic Viscosity		
°C	°F	g/cm^3	kg/m^3	$slugs/ft^3$	$dyn \cdot s/cm^2$ (poise)	$Pa \cdot s$ ($N \cdot s/m^2$)	$lb_f \cdot s/ft^2$	cm^2/s (stoke)	m^2/s	ft^2/s
0	32	0.99987	999.87	1.940	1.794×10^{-2}	1.794×10^{-3}	3.746×10^{-5}	1.794×10^{-2}	1.794×10^{-6}	1.930×10^{-5}
4	39	1.00000	1000.00	1.941	1.568	1.568	3.274	1.568	1.568	1.687
5	41	0.99999	999.99	1.941	1.519	1.519	3.172	1.519	1.519	1.634
10	50	0.99973	999.73	1.940	1.310	1.310	2.735	1.310	1.310	1.407
15	59	0.99913	999.13	1.940	1.145	1.145	2.391	1.146	1.146	1.233
20	68	0.998	998.00	1.937	1.009	1.009	2.107	1.011	1.011	1.088
30	86	0.996	996.00	1.932	0.800	0.800	1.670	0.803	0.803	0.864
40	104	0.992	992.00	1.925	0.654	0.654	1.366	0.659	0.659	0.709
50	122	0.988	988.00	1.917	0.549	0.549	1.146	0.556	0.556	0.598
60	140	0.983	983.00	1.907	0.470	0.470	0.981	0.478	0.478	0.514
70	158	0.978	978.00	1.897	0.407	0.407	0.850	0.416	0.416	0.448
80	176	0.972	972.00	1.885	0.357	0.357	0.745	0.367	0.367	0.395
90	194	0.965	965.00	1.872	0.317	0.317	0.662	0.328	0.328	0.353
100	212	0.958	958.00	1.858	0.284	0.284	0.593	0.296	0.296	0.318

Table B-2 Properties of Air at Atmospheric Pressure

Temperature		Density			Viscosity			Kinematic Viscosity		
°C	°F	g/cm^3	kg/m^3	$slugs/ft^3$	$dyn \cdot s/cm^2$ (poise)	$Pa \cdot s$ $(N \cdot s/m^2)$	$lb_f \cdot s/ft^2$	cm^2/s (stoke)	m^2/s	ft^2/s
0	32	1.293×10^{-3}	1.293	2.510×10^{-3}	1.709×10^{-4}	1.709×10^{-5}	3.568×10^{-7}	0.1322	1.322×10^{-5}	1.427×10^{-4}
50	122	1.093	1.093	2.122	1.951	1.951	4.074	0.1785	1.785	1.921
100	212	0.946	0.946	1.836	2.175	2.175	4.541	0.2299	2.299	2.474
150	302	0.834	0.834	1.619	2.385	2.385	4.980	0.2860	2.860	3.077
200	392	0.746	0.746	1.448	2.582	2.582	5.391	0.3461	3.461	3.724
250	482	0.675	0.675	1.310	2.770	2.770	5.784	0.4104	4.104	4.416
300	572	0.616	0.616	1.196	2.946	2.946	6.151	0.4782	4.782	5.145
350	662	0.567	0.567	1.101	3.113	6.500	6.500	0.5490	5.490	5.907
400	752	0.525	0.525	1.019	3.277	3.277	6.842	0.6246	6.246	6.721
450	842	0.488	0.488	0.947	3.433	3.433	7.168	0.7035	7.035	7.570
500	932	0.457	0.457	0.887	3.583	3.583	7.481	0.7840	7.840	8.436

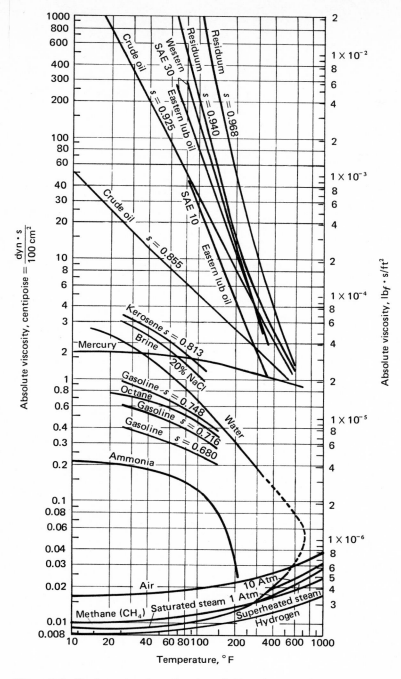

Figure B-1 Absolute viscosity of various fluids, where s refers to the substance relative to that of water at 60°F. (From R. L. Daugherty and A. C. Ingersoll, "Fluid Mechanics," McGraw-Hill Book Company, New York, 1954. Used by permission.)

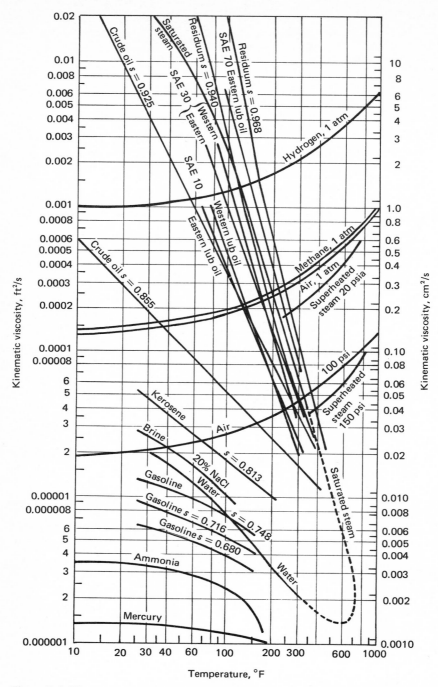

Figure B-2 Kinematic viscosity of various fluids, where *s* has the same meaning as in Fig. B-1. (From R. L. Daugherty and A. C. Ingersoll, "Fluid Mechanics," McGraw-Hill Book Company, New York, 1954. Used by permission.)

UNITS AND DIMENSIONS

Throughout the world there are two main systems of units in current engineering usage, the English system and the metric system. Various versions of both systems are used, but the new SI (metric) system is scheduled to become the universal system in the United States and throughout the world. Currently, both the SI and English systems are used in engineering practice in the United States, but as the English system is phased out, more and more use will be made of the SI system and it is essential to be able to convert between the two systems.

In fluid mechanics we must deal with measurable quantities such as pressure, velocity, density, and viscosity. These quantities are related through equations derived from laws or definitions. Each contains some or all of the basic dimensions of force (F), mass (M), length (L), time (T), and temperature (θ). For quantitative purposes a set of units must be established for these basic dimensions. (In electromagnetic theory there is one additional basic dimension, which is arbitrary. It is often convenient to take this dimension as charge in coulombs or current in amperes.)

Equations expressing relationships among physical quantities must be dimensionally homogeneous. That is, each term of an equation must have the same dimensions.

A difficulty arises when the units of mass and force are established independently. From Newton's law we find that force and mass are not independent dimensions. Force is proportional to the product of mass and acceleration; that is,

$$F \propto ma$$

And by requiring dimensional homogeneity, we find that

$$F = \frac{ML}{T^2} \quad \text{or} \quad M = \frac{FT^2}{L}$$

The actual choice of basic dimensions, then, is somewhat arbitrary. One may use an F, L, T, θ system or an M, L, T, θ system. Then all other dimensions may be expressed in terms of the chosen independent basic dimensions by means of laws and definitions.

Now we may use Newton's law to define the unit of mass in terms of force and acceleration. We write

$$F = ma \quad \text{or} \quad m = \frac{F}{a}$$

If the unit of force is lb_f and the unit of acceleration is ft/s^2, then the unit of mass is

$$\frac{lb_f \cdot s^2}{ft}$$

This is the amount of mass that is accelerated at the rate of 1 ft/s^2 when acted upon by 1 lb_f. This unit of mass is called a *slug*.

Another unit of mass established independently of Newton's law is also used. The pound mass (lb_m) is defined as the amount of mass that would be attracted toward the Earth's surface by a force of 1 lb (at a particular location). When units of force and mass are defined independent of Newton's law, we must write the equation with a conversion factor, k, in order to make the equation dimensionally homogeneous. We have

$$F = kma \quad \text{or} \quad F = \frac{1}{g_c} ma$$

and

$$g_c = \frac{ma}{F}$$

The quantity g_c has a numerical value and units that depend on the particular units chosen for force, mass, and acceleration. Some particular sets of values are

$$g_c = \frac{1(\text{slug})(\text{ft/s}^2)}{\text{lb}_f} \qquad g_c = \frac{1(\text{g})(\text{cm/s}^2)}{\text{dyn}} \qquad g_c = \frac{980(\text{g}_m)(\text{cm/s}^2)}{\text{g}_f}$$

$$g_c = \frac{32.2(\text{lb}_m)(\text{ft/s}^2)}{\text{lb}_f} \qquad g_c = \frac{1(\text{kg})(\text{m/s}^2)}{\text{N}} \qquad g_c = \frac{9.8(\text{kg}_m)(\text{m/s}^2)}{\text{kg}_f}$$

where we see that

$$1 \text{ slug} = 32.2 \text{ lb}_m$$

and we note that 1 lb_m will be accelerated at the rate of 32.2 ft/s^2 when acted upon by 1 lb_f. The slug in English units and the kilogram in SI units are the units of mass implied throughout this book, and thus g_c does not appear in the equations.

In summary, it is useful to write the equation $F = ma$ in various systems of units. The following equations indicate the relationships among the dimensions and units.

$$\text{CGS system:} \quad F\,(\text{dyn}) \qquad = m\,(\text{g}) \quad \times a\,(\text{cm/s}^2)$$

$$\text{SI (MKS) system:} \quad F\,(\text{N}) \qquad = m\,(\text{kg}) \quad \times a\,(\text{m/s}^2)$$

$$\text{FPS system:} \quad F\,(\text{poundals}) = m\,(\text{lb}_m) \quad \times a\,(\text{ft/s}^2)$$

$$\text{FSS system:} \quad F\,(\text{lb}_f) \qquad = m\,(\text{slugs}) \times a\,(\text{ft/s}^2)$$

Table C-1 Dimensions and Units: Mechanics

Physical Quantity	Dimensions		Units			
			Metric		English	
	MLT System	FLT System	CGS System	SI System	FPS System	Engineering System
Length	L	L	cm	m	ft	ft
Mass	M	$FL^{-1}T^2$	g	kg	lb_m	slug
Time	T	T	s	s	s	s
Velocity	LT^{-1}	LT^{-1}	cm/s	m/s	ft/s	ft/s
Acceleration	LT^{-2}	LT^{-2}	cm/s²	m/s²	ft/s²	ft/s²
Force	MLT^{-2}	F	dyn	N	pdl	lb_f
Momentum, impulse	MLT^{-1}	FT	g·cm/s = dyn·s	N·s	pdl·s	lb_f·s
Energy, work	ML^2T^{-2}	FL	dyn·cm = erg	N·m = J	ft·pdl	ft·lb_f
Power	ML^2T^{-3}	FLT^{-1}	dyn·cm/s = erg/s	J/s = W	ft·pdl/s	ft·lb_f/s
Density	ML^{-3}	$FL^{-4}T^2$	g/cm³	kg/m³	lb_m/ft³	slug/ft³
Angular velocity	T^{-1}	T^{-1}	rad/s	rad/s	rad/s	rad/s
Angular acceleration	T^{-2}	T^{-2}	rad/s²	rad/s²	rad/s²	rad/s²
Torque	ML^2T^{-2}	FL	dyn·cm	N·m	ft·pdl	ft·lb_f
Angular momentum	ML^2T^{-1}	FLT	g·cm²/s	kg·m²/s	lb_m·ft²/s	slug·ft²/s
Moment of inertia	ML^2	FLT^2	g·cm²	kg·m²	lb_m·ft²	slug·ft²
Pressure, stress	$ML^{-1}T^{-2}$	FL^{-2}	dyn/cm²	Pa = N/m²*	pdl/ft²	lb_f/ft²
Viscosity (μ)	$ML^{-1}T^{-1}$	$FL^{-2}T$	dyn·s/cm²	Pa·s = N·s/m²	pdl·s/ft²	lb_f·s/ft²
Kinematic viscosity (v)	L^2T^{-1}	L^2T^{-1}	cm²/s	m²/s	ft²/s	ft²/s
Surface tension	MT^{-2}	FL^{-1}	dyn/cm	N/m	pdl/ft	lb_f/ft

*The unit of kilopascal (kPa) is commonly used.

Table C-2 Dimensions and Units: Heat

	SI (Metric)	English
Temperature	K	°F or °R
Quantity of heat	J	Btu[†]
Heat flow rate	W	Btu/h or Btu/s
Density of heat flow	W/m²	Btu/ft² · h or Btu/ft² · s
Thermal conductivity	W/m · K	Btu/ft · °F · h
Coefficient of heat transfer	W/m² · K	Btu/ft² · °F · h
Heat capacity	J/K	Btu/°F
Specific heat capacity	J/kg · K	Btu/lb$_m$ · °F or Btu/slug · °F
Specific energy	J/kg or kJ/kg*	Btu/lb$_m$ or Btu/slug
Specific enthalpy	J/kg or kJ/kg*	Btu/lb$_m$ or Btu/slug

*kilojoules per kilogram, which is a more convenient unit.

[†] Btu or ft · lb$_f$ may be used as convenient. The conversion is 778 ft · lb$_f$/Btu.

Table C-3 Conversion Factors

Length	1 kilometer (km)	$= 1000$ meters
	1 meter (m)	$= 100$ centimeters
	1 centimeter (cm)	$= 10^{-2}$ m
	1 millimeter (mm)	$= 10^{-3}$ m
	1 micron (μ)	$= 10^{-6}$ m
	1 millimicron (mμ)	$= 10^{-9}$ m
	1 angstrom (A)	$= 10^{-10}$ m
	1 inch (in.)	$= 2.540$ cm
	1 foot (ft)	$= 30.48$ cm
	1 mile (mi)	$= 1.609$ km
	1 mil	$= 10^{-3}$ in.
	1 centimeter	$= 0.3937$ in.
	1 meter	$= 39.37$ in.
	1 kilometer	$= 0.6214$ mile

Area 1 square meter (m²) $= 10.76$ ft²
1 square foot (ft²) $= 929$ cm²

Volume 1 liter (*l*) $= 1000$ cm³ $= 1.057$ quart (qt) $= 61.02$ in.³ $= 0.03532$ ft³
1 cubic meter (m³) $= 1000$ *l* $= 35.32$ ft³
1 cubic foot (ft³) $= 7.481$ U.S. gal $= 0.02832$ m³ $= 28.32$ l
1 U.S. gallon (gal) $= 231$ in.³ $= 3.785$ *l*
1 British gallon $= 1.201$ U.S. gallon $= 277.4$ in.³

Mass 1 kilogram (kg) $= 2.2046$ lb$_m$ $= 0.06852$ slug
1 lb$_m$ $= 453.6$ g $= 0.03108$ slug
1 slug $= 32.174$ lb$_m$ $= 14.59$ kg

Table C-3 Conversion Factors (*Continued*)

Speed	1 km/h = 0.2778 m/s = 0.6214 mi/h = 0.9113 ft/s
	1 mi/h = 1.467 ft/s = 1.609 km/h = 0.4470 m/s
Density	1 g/cm^3 = 10^3 kg/m^3 = 62.43 lb$_m$/ft^3 = 1.940 slug/ft^3
	1 lb$_m$/ft^3 = 0.01602 g/cm^3
	1 slug/ft^3 = 0.5154 g/cm^3
Force	1 newton (N) = 10^5 dynes = 0.1020 kg$_f$ = 0.2248 lb$_f$ ·
	1 pound force (lb$_f$) = 4.448 N = 0.4536 kg$_f$ = 32.17 poundals
	1 kilogram force (kg$_f$) = 2.205 lb$_f$ = 9.807 N
	1 U.S. short ton = 2000 lb$_f$
	1 long ton = 2240 lb$_f$
	1 metric ton = 2205 lb$_f$
Energy	1 joule (J) = 1 N · m = 10^7 ergs = 0.7376 ft · lb$_f$ = 0.2389 cal = 9.481 × 10^{-4} Btu
	1 ft · lb$_f$ = 1.356 J = 0.3239 cal = 1.285 × 10^{-3} Btu
	1 calorie (cal) = 4.186 J = 3.087 ft · lb$_f$ = 3.968 × 10^{-3} Btu
	1 Btu = 778 ft · lb$_f$ = 1055 J = 0.293 W · h
	1 kilowatt hour (kW · h) = 3.60 × 10^6 J = 860.0 kcal = 3413 Btu
	1 electron volt (eV) = 1.602 × 10^{-19} J
Power	1 watt (W) = 1 J/s = 10^7 ergs/s = 0.2389 cal/s
	1 horsepower (hp) = 550 ft · lb$_f$/s = 33,000 ft · lb$_f$/min = 745.7 W
	1 kilowatt (kW) = 1.341 hp = 737.6 ft · lb$_f$/s = 0.9483 Btu/s
Pressure	1 N/m^2 = 1 Pascal (Pa) = 10 dyn/cm^2 = 9.869 × 10^{-6} atm = 2.089 × 10^{-2} lb$_f$/ft^2
	1 lb$_f$/in.^{2} = 6895 N/m^2 = 5.171 cm mercury = 27.68 in. water
	1 atmosphere (atm) = 1.013 × 10^5 N/m^2 = 1.013 × 10^6 dyn/cm^2 = 14.70 lb$_f$/in.^{2} = 76 cm mercury = 406.8 in. water

FRICTION FACTOR CURVES

A curve of the friction factor f versus the Reynolds number Re may be determined experimentally. One such curve, known as the Moody diagram, is shown below. (L. F. Moody, Friction Factors for Pipe Flow, *Trans. ASME*, vol. 66, no. 8, p. 671, 1944.)

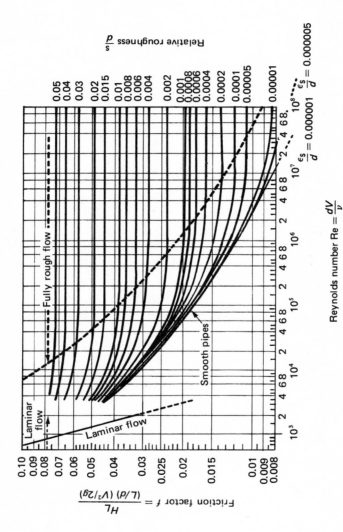

Figure D-1

214

INDEX